HOW TO EVALUATE PROGRESS
IN PROBLEM SOLVING

HOW TO EVALUATE PROGRESS IN PROBLEM SOLVING

Randall Charles
Illinois State University

Frank Lester
Indiana University

Phares O'Daffer
Illinois State University

National Council of Teachers of Mathematics

Sixth printing 1997

Library of Congress Cataloging in Publication Data:

Charles, Randall (Randall I.)
How to evaluate progress in problem solving.

Bibliography: p.
1. Problem solving—Study and teaching. 2. Mathemati-
cal—Testing. I. Lester, Frank K. II. O'Daffer,
Phares G. III. Title.
QA63.C48   1987   510'.7   87-7644
ISBN 0-87353-241-4

The publications of the National Council of Teachers of Mathematics present a variety of viewpoints. The views expressed or implied in this publication, unless otherwise noted, should not be interpreted as official positions of the Council.

Printed in the United States of America

Contents

Introduction:
Understanding the Problem

ANY call for significant curricular change poses new challenges for teachers. This is no less true of the recommendation that problem solving be the focus of school mathematics, even though, generally, the idea has been well received by teachers, curriculum developers, and other educators. With an increased emphasis on problem solving comes the challenge of developing new techniques for evaluating the effectiveness of instruction.

This book addresses that challenge by putting the goals of problem-solving evaluation into sharper focus, describing several classroom evaluation techniques, and illustrating how these techniques might be used in practice.

To begin, let us consider a conversation that could have taken place in the teachers' lounge of any elementary or middle school.

Luis *and* Marion *are relaxing in the teachers' lounge during a brief break;* Linda *enters carrying some papers.*

Linda. Help! I'm glad you two are here; I need your advice.

Luis. What's the problem, Linda? You're usually able to handle anything that comes up.

1

Linda. Hmm, I'm not so sure. Remember that problem-solving workshop we went to last summer? It really fired me up about making a serious effort to include problem solving in my math activities this year. Well, everything was going fine. I had a good collection of problems, and I thought I had a good grasp of the instructional strategy we were shown in the workshop. And the kids—they really enjoy solving and discussing the problems. But . . .

Marion. There's always a "but." We both know how excited you've been in the past several weeks. But nothing could be as good as you were leading us to believe. Tell us about it. What's the problem?

Linda. OK! Today I gave the class a problem to solve. I told them to show as much of their work as possible because they could get partial credit for a good effort even if their answers were wrong. Now I've got to grade their solutions and I don't know what to do. Here, this is the problem they worked on:

> **PROBLEM OF THE DAY**
>
> Marty gets an allowance of $1.50 every Friday. One Friday Marty was given 10 coins in nickels, dimes, and quarters. How many of each kind of coin did Marty get?

Luis. My fifth graders would have a lot of trouble with this problem. There are too many pieces of information to deal with.

Marion. Yeah, I agree, but if they had some coins it might be OK for them.

Linda. Actually, the problem wasn't that difficult for my class, and they enjoyed solving it. Some of the students used play money. Others tried to make tables to organize their work. I like both approaches because we have solved several problems by using aids or making tables. But look at these two papers [figs. 1 and 2] and tell me what you think of their work:

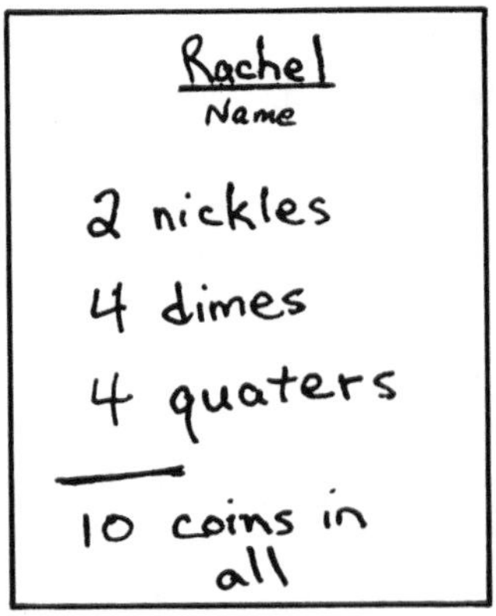
Rachel
Name

2 nickles
4 dimes
4 quaters

10 coins in all

Fig. 1

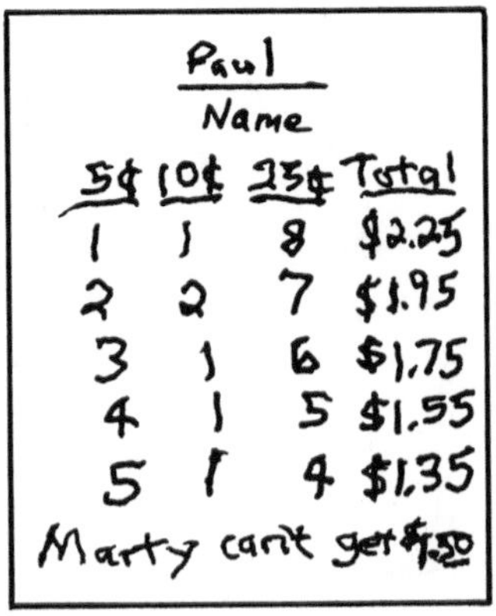
Paul
Name

5¢	10¢	25¢	Total
1	1	8	$2.25
2	2	7	$1.95
3	1	6	$1.75
4	1	5	$1.55
5	1	4	$1.35

Marty can't get $1.50

Fig. 2

Marion. Let me see. What's the correct answer? Is Rachel's answer correct?

Luis. It is, but I think I see what Linda's problem is. Rachel got the answer, but it's impossible to tell how she got it. In fact, she may have simply found three numbers that add up to ten without even considering if that combination of coins totals $1.50.

Marion. I see what you mean now, but she did get it right so she should get credit.

Linda. Maybe! Take a look at Paul's paper. His answer is wrong but he did several things that really impress me. He used all the important information, he made a good, nicely labeled table, and he was willing to accept that the problem might not have an answer. Also, notice that his entries in the table are systematically organized.

Luis. Your point is that somehow you want to give him credit for a good effort.

Linda. That's part of it, but there's more to it . . .

Marion. Look, it's simple. Give Rachel an A, or whatever you put on correct work, and give Paul an A−.

Linda. It may be simple for you but I want to give Paul more specific feedback than that. I want him to know that he did several things well but he didn't quite do all that was needed. Also, I'm worried about Rachel. I don't want to encourage her to make guesses without checking them. I think giving her an A might do just that.

Luis. What you need are some specific guidelines for evaluation. I wonder if any evaluation guidelines have been written?

You may have seen the picture of a bedraggled young teacher, loaded down with papers and books, leaving her classroom after a long day. The caption reads: "Nobody said it was going to be easy!" To some extent, this sentiment holds for teaching problem solving. In the conversation above, Linda indicates that she was, and still is, enthusiastic about teaching problem solving. However, she has come up against an unexpected stumbling block. Her fellow teachers, Luis and Marion, are supportive but of little help to her. Linda needs some specific ideas and guidelines for evaluating her students' progress. This booklet provides such a set of guidelines.

3

CURRENT STATUS OF MATHEMATICAL PROBLEM-SOLVING EVALUATION

A few years ago one of the authors made a presentation on the role of problem solving in school mathematics to a group of state and district mathematics curriculum coordinators and supervisors. During the discussion session that followed, one perceptive educator insisted, "I like what you're saying, but none of your ideas have a chance of succeeding unless teachers change their evaluation methods and states change their testing programs." This proclamation sparked one of the liveliest discussions about mathematical problem solving the speaker had ever been involved in. More importantly, from the comment and the discussion that ensued, we concluded that it was time to begin taking a serious look at current evaluation practices.

Three important conclusions resulted from our investigation: (1) answer-focused paper-and-pencil tests are by far the most common type of assessment of mathematical problem-solving progress; (2) some teachers, fewer school districts, and even fewer states have begun to adopt a process-oriented view of problem-solving evaluation; and (3) several evaluation techniques developed for use in areas unrelated to mathematics suggest ways to measure a wide range of processes, skills, and attitudes associated with mathematical problem solving. This book has grown out of our investigation. It represents our response to Recommendation 5 of the NCTM's *Agenda for Action* that "the success of mathematics programs and student learning [must] be evaluated by a wider range of measures than conventional testing" (NCTM 1980, p. 1).

HOW TO USE THIS BOOK

It is our intention to help you develop both an understanding of, and skill in performing, problem-solving evaluation through examples of practical classroom situations for which some type of evaluation is desirable. At the same time, this is much more than a collection of classroom episodes. In fact, the primary focus is on specific ideas and guidelines for evaluating the progress your students are making toward becoming good problem solvers.

As you know, problem solving is a complex form of intellectual activity. It is because of this complexity that you should develop a good understanding of the nature of problem solving and various evaluation techniques associated with it before you attempt to set up your own evaluation program. To help you gain this perspective, we have organized the booklet around four key questions associated with classroom problem-solving evaluation. These four questions and a very brief description of each are given in figure 3.

Before you can establish a good evaluation program, you must have a very clear notion about what problem solving involves and what the goals of problem-solving instruction should be. Consequently, we urge you to con-

How This Book Is Organized

Key Questions	Discussion Description
Question 1: What are you trying to evaluate?	The skills and abilities involved in problem solving and the goals for teaching problem solving are considered.
Question 2: What are some key evaluation techniques?	Four major categories of evaluation techniques, ranging from observations of students working in small groups to the analysis of students' written work, are discussed.
Question 3: How do you design and manage an evaluation program?	Setting up an evaluation program should be based on a set of essential guidelines. Sample evaluation programs are described.
Question 4: How do you use evaluation results?	Evaluation data can assist in making decisions about teaching methods, content of instruction, classroom management, and grading. Tips for using evaluation results are given.

Fig. 3

sider the discussion of Question 1 very carefully before continuing on to the other three. However, if you have thought long and hard about the nature and importance of problem solving in the mathematics curriculum, you might choose to skim the discussion of Question 1 and go on to the specific evaluation techniques.

We are strong believers in "learning by doing." For this reason we urge you to take an *active* part in studying the material we present. In particular, we think it is essential that you take seriously the "Give it a try" sections that accompany each evaluation technique. When you come to these sections, take the opportunity to practice that particular technique before trying it out in your own classroom. By so doing, you should find it easier to develop a clearer perspective as to the potential usefulness of the technique.

Enough said! Let's begin!

1

What Are You Trying to Evaluate?

YOUR plan for evaluating progress in problem solving should build on the goals you select. In this section, we discuss several of the most important goals for teaching problem solving. In the next section we describe some evaluation techniques you can use to assess these goals.

Before discussing specific goals, it may be helpful to think briefly about what is involved in mathematical problem solving. Problem solving is an extremely complex activity. It involves the recall of facts, the use of a variety of skills and procedures, the ability to evaluate one's own thinking and progress while solving problems, and many other capabilities. Furthermore, success in problem solving very much depends on the student's interest, motivation, and self-confidence. In short, solving problems involves the coordination of knowledge, previous experience, intuition, attitudes, beliefs, and various abilities. Here are seven goals we have identified for teaching problem solving (their order does not reflect their relative importance):

1. To develop students' *thinking skills*
2. To develop students' abilities to *select and use problem-solving strategies*
3. To develop *helpful attitudes and beliefs* about problem solving
4. To develop students' abilities to *use related knowledge*
5. To develop students' abilities to *monitor and evaluate their thinking* and progress while solving problems
6. To develop students' abilities to *solve problems in cooperative learning situations*
7. To develop students' abilities to *find correct answers* to a variety of types of problems

Let us consider these goals individually.

Goal 1: To develop students' problem-solving *thinking skills*

Most people agree that the process of solving problems involves a variety of thinking skills. In our study of the literature we have found seven thinking skills to be particularly important.

1. *Understand /formulate the question in a problem.* One of the first tasks in solving a problem is to find or formulate the question and to "make sense" of it. For different problems, the question may appear in different places.

7

Occasionally, a question appears as a statement. Making sense of the question involves understanding the meaning of specific words in the problem (or in the question itself) and also recognizing how the question relates to other statements in the problem.

2. *Understand the conditions and variables in the problem.* Consider the following problem:

> Tom and Sue saw some chickens and pigs in a barnyard. Tom said, "There are 18 chickens and pigs." Sue said, "Yes, and altogether they have 52 legs." How many chickens and how many pigs did they see?

This problem has two conditions: (1) There are 18 animals in all, and (2) there are 52 legs. Also, there are two variables in this problem: (1) the number of chickens and (2) the number of pigs. During the process of understanding the conditions and variables, the problem solver "internalizes the problem." That is, he or she develops a sense of how the conditions and variables relate to each other and clarifies the meaning of the information explicitly stated or implied in the problem. Often the process of understanding the conditions and variables in a problem is helped by making a model, a diagram or picture, or a list of key ideas arranged in a particular way.

3. *Select or find the data needed to solve the problem.* Several important problem-solving processes are suggested by this general statement. A problem solver must be able to identify needed data, eliminate data not needed, and collect and use data from a variety of sources such as graphs, maps, or tables. Data-selection processes are closely connected to processes involved in understanding the question, conditions, and variables in a problem.

4. *Formulate subproblems and select appropriate solution strategies to pursue.* This is the planning phase. Here the problem solver must determine if there are subproblems or subgoals to be solved and which solution strategies might be tried. To solve multiple-step problems or process problems, the problem solver often must identify subgoals to be reached as progress is made toward the solution. He or she must also decide which strategy or strategies to use. It is one thing to know *how* to use particular solution strategies and quite another thing to know *when* to use them. For example, a student may know how to multiply but not know when to multiply. Similarly, a student might know how to find a pattern but not know when to look for a pattern in solving problems. This thinking skill involves a decision as to which strategy or strategies to try. (A listing of problem-solving strategies is given in the discussion for goal 2.)

To examine the process of selecting subproblems and a solution strategy, consider the following problem:

8

> Marty was broke when he received his weekly allowance on Monday. On Tuesday, he spends $1.25 of it. On Wednesday, his sister pays him the $1 she owes him. How much is Marty's allowance if he now has $2.25?

To solve this problem, you might reason as follows: "I know Marty has $2.25 now. First, I'll find out how much he had before his sister paid him $1. Then I'll find out how much he had before he spent $1.25. This will give me the answer." There are two subgoals here—to (1) determine how much Marty had before his sister paid him $1 and (2) determine how much he had before spending $1.25. The strategy used is called "work backward." This example illustrates that subproblems or subgoals are intermediate stages along the way toward a solution that the problem solver consciously tries to reach. The solution strategy is the plan of attack for reaching the subgoals and ultimately the solution.

5. *Correctly implement the solution strategy or strategies and solve subproblems.* As mentioned above, the problem solver must know *how* to implement solution strategies. Implementing a strategy may involve being able to perform computations, use logical reasoning, or solve equations, but it may also involve activities such as making a table, making a list, and so on. Similarly, after identifying and ordering subgoals, the problem solver must be able to attain them.

6. *Give an answer in terms of the data in the problem.* The problem solver must be able to give an answer in terms of the relevant features of the problem. This may mean giving the correct unit to accompany the numerical part of an answer or stating the answer in a complete sentence. For example, in the "chickens and pigs" problem stated earlier, the student should be able to state the answer in terms of chickens and pigs (10 chickens and 8 pigs, not simply 10 and 8).

7. *Evaluate the reasonableness of the answer.* The problem solver should be able to determine whether or not the answer makes sense. This process might involve rereading the problem and checking the answer against the relevant information (conditions and variables) and the question. Students might also use various estimation techniques to determine if an answer is reasonable.

Goal 2: To develop students' abilities to *select and use problem-solving strategies*

Students' confidence and abilities are greatly enhanced when they have mastered a repertoire of *strategies* for solving problems. Beginning in kindergarten, students should be taught strategies for solving problems. Instruction should build on the problem-solving techniques many children use

9

naturally and bring with them when they start school. As students move through the grades, their skill and understanding with strategies can be enhanced and deepened, and new, more sophisticated strategies can be introduced. Although names different from those given below can be used, the following strategies are among those that might be introduced in a problem-solving program.

- Guess, check, revise.
- Draw a picture.
- Act out the problem.
- Use objects.
- Choose an operation(s).
- Solve a simpler problem.
- Make a table.
- Look for a pattern.
- Make an organized list.
- Write an equation.
- Use logical reasoning.
- Work backward.

Goal 3: To develop *helpful attitudes and beliefs* about problem solving

Students' attitudes and beliefs about problem solving and about themselves can influence their performance greatly. Attitudes and beliefs can be helpful or debilitating. Examples of debilitating attitudes and beliefs are, "All problems can be solved in only one way"; "If I can't get the answer right away, I'll never get it." Examples of helpful attitudes and beliefs are, "Many problems can be solved in more than one way"; "Many problems have more than one answer"; "If the first strategy I try doesn't help, I'll try to find one that might help." Instructional programs for problem solving should foster the development of helpful attitudes and beliefs and dispel erroneous and debilitating ones.

Goal 4: To develop students' abilities to *use related knowledge*

Successful problem-solving performance is often influenced by the student's ability to accurately recall and apply specific knowledge. This knowledge might be mathematical (for example, how to find a least common multiple) or relevant to the context of the problem. (For instance, a problem about baseball might require knowing there are nine positions.) Students need to be taught *how* to use specific mathematical knowledge and *when* to use that knowledge. Furthermore, they need to practice applying newly learned mathematical skills in problem-solving situations.

Goal 5: To develop students' abilities to *monitor and evaluate their thinking* and progress while solving problems

Too many students adopt what initially seems like a reasonable approach to a problem and then proceed to follow it without further evaluation of their decision as long as they are able to keep working. They need to learn that it is valuable to slow down periodically and reflect on what they are trying to do, what they have done, and what they still need to do. An instructional program for problem solving should develop in students the skills needed for

10

monitoring and evaluating their thinking and progress while they are solving problems. The program should also help students make these monitoring and evaluation skills part of their consciousness.

Goal 6: To develop students' abilities to *solve problems in cooperative learning situations*

At one level, this instructional goal is concerned with the development of helpful social skills involved in cooperative learning situations. However, it is also concerned with many kinds of intellectual skills that are needed for successful problem solving in groups. Such skills as clarifying one's ideas, evaluating another's ideas, and comparing alternatives promote successful problem solving and are developed best in cooperative situations.

Goal 7: To develop students' abilities to *find correct answers* to different types of problems

The student who masters goals 1 through 6 but not goal 7 would certainly be deficient at problem solving. After all, the reason we try to solve a problem is to get the correct answer. An important goal for problem-solving instruction is for students to gain the ability to find correct answers to problems. However, in the classroom, teaching and evaluation should focus not only on their ability to get correct answers but also on the abilities reflected by goals 1 through 6.

An instructional program for problem solving should provide experience in solving a variety of problems. We believe there are at least four types of problems that should be used in a problem-solving program: one-step problems, multiple-step problems, process problems, and applied problems. Let us examine each of them.

One-step problems. One-step problems provide experience in translating story situations to a number sentence involving addition, subtraction, multiplication, or division. These are the familiar "story problems" that have always been part of school mathematics programs. The primary strategy used to solve one-step problems is *choose the operation.*

Sample (grade 2):

The sales record for second-grade students was 75 boxes. The sales record for first-grade students was 56 boxes. How many more boxes did the second-grade students sell than the first-grade students?

Solution:

Possible solution strategy: Choose the correct ration.

$$75 - 56 = 19$$

The first grade sold 19 fewer boxes than the second grade.

11

Multiple-step problems. The difference between one-step and multiple-step problems is in the number of operations needed to find a solution. The primary strategy used to solve multiple-step problems is *choose the operations.*

Sample (grade 4):

Carrie is allowed to watch 35 hours of television each week. If she watches 20 hours on the weekend, how many hours, on the average, can she watch television each weekday?

Solution:

Possible solution strategy: Choose the correct operations.

$$35 - 20 = 15 \qquad 15 \div 5 = 3$$

Carrie can watch an average of 3 hours of television each weekday.

Process problems. Process problems are solved using such strategies as guess-check-revise, draw a picture, make a table, look for a pattern, work backward, solve a simpler problem, and make an organized list. The solution may involve some computation, but these problems cannot be solved by the student by simply choosing the operation or operations.

Sample (grade 6):

Six people entered a tennis tournament. Each player played each other person one time. How many games were played?

Solution:

Possible solution strategies:

Make an organized list.
Look for a pattern.

A	B	C	D	E	F
B	C	D	E	F	
C	D	E	F		
D	E	F			
E	F				
F					

$$5 + 4 + 3 + 2 + 1 + 0 = 15$$

15 games were played.

Applied problems. Applied problems (sometimes called situational problems) require the problem solver to collect data outside the problem statement. Many applied problems require the student to formulate a clear statement of the problem and subproblems and identify assumptions that need to be made and the data needed for finding a solution.

Sample (grade 3):

You and a friend decide to operate a lemonade stand. You need to decide what to charge for each glass of lemonade.

Some comments on the goals for teaching problem solving

The relationships among the seven goals are important to consider. In particular, the relationship between goals 1 through 6 and goal 7 has implications for the evaluation plan you develop. A sports analogy may be helpful. A professional golfer usually has a beautiful golf swing. To have a good swing, the golfer must master a number of subskills. For example, the club must be positioned in the hands correctly, the feet must be in a proper position at all times, and the arms should be in a certain position at the completion of the backswing. Many golfers can perform each of these subskills and still not hit the ball with anywhere near the quality of the professional. Why? Perhaps the reason is obvious. A successful golf swing requires not only the mastery of the subskills involved but also the coordination of those skills. The ability to coordinate the skills is more difficult to master than the subskills and, as a result, takes longer to achieve.

The relationship of the first six goals with the seventh is similar to the relationship of the subskills of a golf swing to the coordination of those skills in successfully hitting the ball. The ability to coordinate skills and abilities for problem solving develops gradually over time. This means that the ability to get correct answers will develop more slowly than the skills and abilities associated with goals 1 through 6. An evaluation plan for problem solving should not focus exclusively on the assessment of goal 7. Such a plan could not provide information about a student's strengths and weaknesses, information needed to make instructional decisions regarding the content, or information about methods of teaching problem solving. We should look for progress within and across years in each of the goals given above and not rely on performance in any one of these as a sole measure of success or failure.

2

What Are Some Evaluation Techniques?

THE goals for teaching problem solving presented in the preceding section suggest that it is important to evaluate students' progress in problem solving in two major areas: (1) *performance* in using a variety of problem-solving skills and strategies, and (2) *attitudes and beliefs* regarding problem solving. This section describes several techniques for evaluating these two important, interrelated outcomes. Included are techniques for—

1. observing and questioning students;
2. using assessment data from students;
3. using holistic scoring techniques;
4. using multiple-choice and completion tests.

Your choice of evaluation techniques might be based on such factors as (1) the type of problem-solving skill or outcome being measured, (2) the number of students being evaluated, (3) the time available for evaluation, (4) your experience in teaching and evaluating problem solving, (5) how you intend to use the results of the evaluation, and (6) the availability of evaluation materials. Therefore, your task is to select those techniques that can best help you measure the goals you choose to emphasize and are most feasible for you to use. A later chapter, "How Do You Organize and Manage an Evaluation Program," identifies some typical ways teachers have selected and used the techniques described here to build an assessment plan for problem solving.

The discussion of each technique, when appropriate, will be organized according to the following outline:

- What is it?
- Give it a try!
- What are its advantages?
- What are its disadvantages?
- When should it be used?
- How does one develop it?

Where appropriate, we give a sample of a technique and an opportunity for you to develop your evaluation skills by using it. Remember that these techniques are only samples that you can modify to meet your own needs or preferences.

TECHNIQUE 1: OBSERVING AND QUESTIONING

Observing and questioning students while they solve problems can yield valuable information about their performance, attitudes, and beliefs. Observations and questions can be handled informally as you move about the room while they are working, or formally through structured individual interviews. This section discusses both informal and formal techniques for observing and questioning students.

Direct observation and careful questioning of students as they solve problems are among the best methods of evaluating some of the goals of problem solving. A thorough evaluation cannot be carried out if methods are limited to the analysis of written work. When observing and questioning, it is important to make an objective record of the student's responses regarding the skill or attitude you are evaluating. The techniques described in this section vary according to the degree of formality of the approach, the setting in which the evaluation is done, and the type of recording devices used.

One approach is the *informal observation and questioning* of an individual, a small group, or a class. Another approach is a *structured interview*. Let's now describe these two methods in more detail.

Informal Observation and Questioning

What is informal observation and questioning?

In this approach, an evaluator observes an individual, small group, or class solving problems and while doing so asks informal evaluative questions and records observations. The method can be used to assess both performance and attitudes and beliefs. Techniques for observing, questioning, and recording are discussed below.

Observation techniques. You can learn a lot about students' problem-solving performance and attitudes simply by observing them in a problem-solving situation. First, it is important that they learn not to be distracted by your presence. If students work in small groups, you can unobtrusively move among them and observe how they work with each other on a problem. Second, your observations should be focused. Limit your observations by looking for those aspects of performance and attitude that cannot be more easily evaluated using other techniques. Select only a few students at a time to observe, and decide ahead of time which aspects of their problem-solving behavior you will concentrate on. Finally, even though it is advisable to have an observation plan, it is also important to be flexible enough to note other significant behavior and to question students, if appropriate, to gain deeper insight.

Questioning techniques. There are different purposes for asking questions in the classroom. One is to stimulate mathematical thinking. Another is to

16

help the student solve a problem. However, the purpose of the questioning techniques described here is to help the questioner evaluate the student's problem-solving skills and attitudes. Thus questions that shed light on student's processes or feelings about problem solving should be chosen. These questions might take some of the following forms:

1. How did you . . . ?
2. Why did you . . . ?
3. What did you try . . . ?
4. How do you know that . . . ?
5. Have you . . . ?
6. How did you happen to . . . ?
7. How did you decide whether . . . ?
8. Can you describe . . . ?
9. Are you sure that . . . ?
10. What do you think . . . ?
11. How do you feel about . . . ?

For example, as Sue considers a problem, you might gain insight into her ability to understand the problem and look for the question by asking "*What did you do first* when you started to solve the problem?" or "*What do you think is most important* in trying to understand a problem?" As she continues to think about the information in the problem, you might assess her ability to deal with data by asking such questions as "*How do you know that* (a certain piece of information) is needed to solve the problem?" While Sue is solving the problem, you might investigate her ability to choose appropriate strategies by asking a question such as "*Have you* used any strategies in solving the problem? Which ones?" As she proceeds to find the answer, you could ask questions such as "*How did you decide whether* to multiply or add to find the answer?" When she gets an answer, you could ask, "*Are you sure* this is the answer to the question asked in the problem? Why?" Finally, after the problem is solved, it may be useful to ask, "*Can you describe* your solution to the problem?" or "*How do you feel* about your experience with this problem?"

Recording techniques. As you observe and question students in a problem-solving situation, record your findings briefly and objectively. This record should be made as soon as possible after the observation, and can include both a description of the situation and your interpretations of it. Methods for recording observations include using a comment card (fig. 4), a checklist (fig. 5), or a rating scale (fig. 6).

Problem-solving Observation Comment Card

Student _____Sue Trent_____   Date ___10/5___

Comments:

Knows how and when to look for a pattern.
Knows that a table will help her find a pattern.
Keeps trying even when she has trouble
finding a solution.
Needs to be reminded to check her solutions.

Fig. 4

Problem-solving Observation Checklist

Student _____   Date _____

____ 1. Likes to solve problems
____ 2. Works cooperatively with others in the group
____ 3. Contributes ideas to group problem solving
____ 4. Perseveres—sticks with a problem
____ 5. Tries to understand what a problem is about
____ 6. Can deal with data in solving problems
____ 7. Thinks about which strategies might help
____ 8. Is flexible—tries different strategies if needed
____ 9. Checks solutions
____ 10. Can describe or analyze a solution

Fig. 5

Give it a try!

Present the following problem to a student or friend:

> A frog is at the bottom of a 10-meter well. On the first day it climbs up 5 meters, but at night it slips back 4 meters. If it does this each day, on what day will it get out of the well?

Use the techniques described above to observe and question the person as he or she attempts to solve the problem. Record the results.

18

Problem-solving Observation Rating Scale

Student _____ Date _____

	Frequently	Sometimes	Never
1. Selects appropriate solution strategies	___	___	___
2. Accurately implements solution strategies	___	___	___
3. Tries a different solution strategy when stuck (without help from the teacher)	___	___	___
4. Approaches problems in a systematic manner (clarifies the question, identifies needed data, plans, solves, and checks)	___	___	___
5. Shows a willingness to try problems	___	___	___
6. Demonstrates self-confidence	___	___	___
7. Perseveres in problem-solving attempts	___	___	___

Fig. 6

What are some advantages of informal observation and questioning?

Advantages of this technique include the following:

- It allows for evaluation in a natural classroom problem-solving setting.
- It is flexible, allowing for evaluation of only a few students at a time.
- It allows for evaluation focused on limited, specific aspects of student behavior.
- It allows for evaluation of aspects of performance and attitude that are difficult, if not impossible, to evaluate using other techniques.
- It provides a record of observed growth in the development of specific problem-solving skills and attitudes and a check on evaluations using other methods.

What are some disadvantages of informal observation and questioning?

Disadvantages include the following:

- It may interfere with other important management and instructional responsibilities.
- It is time- and thought-consuming to evaluate all students regularly in this manner and keep appropriate records.
- It requires considerable insight into problem solving to choose the most appropriate questions to ask and processes to evaluate.
- It needs to be carefully planned to give information beyond what can be collected by evaluation of students' written work.
- It is difficult to be unbiased when observing student responses.

When and how should informal observation and questioning be used?

Informal observation and questioning is useful for evaluating some of the important goals we have suggested for problem-solving performance and attitudes. For example, this technique may be the most useful one available for evaluating a student's thinking processes during a problem-solving session. Also, it is an excellent way to assess a variety of attitudes and beliefs, including a willingness to try problems and perseverance in solving problems. Observation is also an effective means of evaluating the students' ability to work cooperatively with others in solving problems.

Informal observation and questioning can be used when students are working individually, in small groups, or as a whole class. It is probably most effective during individual or small-group work, since a teacher has limited time to take notes during a whole-class discussion.

Prior to the lesson, choose both the aspect of performance or attitude you wish to evaluate and the students to be evaluated. Any checklists or rating scales should be prepared ahead of time. As the chosen students solve problems, observe them, listen as they talk to others, and pose questions according to what you want to evaluate. Record your observations on the spot if possible, rather than relying on memory. Limit your goals and try not to do too much. *It is not intended that you evaluate every student in every problem-solving experience.* Rather, in a given problem-solving situation, you may wish to focus on one to four students.

How does one develop the checklists and rating scales?

The checklist and the rating scale given earlier are samples. They may be revised or used as guides to develop other scales that fit your needs. A general procedure for creating them is described as follows:

1. Determine the goal(s) of performance or attitude you wish to evaluate.

2. List specific student actions, thoughts, or attitudes that would indicate the attainment of the goal or goals.

3. Write items on the checklist or rating scale that describe the specifics in step 2. (If a rating scale is to be used, select an appropriate scale.)

Note that a checklist can include several goals, or it can be focused on a single goal or subgoal. Select those goals that cannot be evaluated more easily by other means.

Structured Interviews

What is a structured interview?

This technique involves the observation and questioning of students during a problem-solving session. However, unlike the informal method, a structured interview involves no more than two students and, typically, only one. An interview is made systematic, or structured, by presenting a pre-selected sequence of problems, selecting specific categories in which to ask sequenced probing questions, or by using computer software to present students with problem-solving situations in which standardized questions are asked. Figure 7 shows a sample of one interviewer's plan.

Interview Plan (Fifth Grade)

1. Establish rapport to help the student feel comfortable.

2. Ask the student to "talk about what he/she is doing or thinking" while solving the problem. Point out in a natural way that this will help you understand more about how fifth graders solve problems and enable you to help them become better problem solvers.

3. Hand this problem to the student:

At an amusement park, Jon and his 5 friends decided to take enough roller coaster rides so that each person would take a ride with every other person exactly once. How many rides were taken if only 2 students went on each ride?

4. As the student attempts to understand the problem question and conditions, observe the student and ask questions such as the following, if appropriate:

   a. What did you do first when given the problem? Next?
   b. What question is asked in the problem? What are the important facts, conditions in the problem? Do you need any information not given in the problem?
   c. Is there anything you don't understand about the problem?

5. As the student works on a solution to the problem, remind him/her again to talk about it, and ask questions such as the following, if appropriate:

   a. What strategy are you using? Do you think it will lead to a solution? Have you thought about using other strategies? Which ones?
   b. Where are you having difficulty? What are your ideas about where to go from here?

6. As the student finds an answer to the problem, observe the ways, if any, in which he/she checks the answer and its reasonableness as a solution. Ask questions such as:

   a. Are you sure this is the correct answer to the problem? Why?
   b. Do you think it is important to check your answer? Why?

7. After the student has solved the problem, ask questions such as:

   a. Can you describe the solution to the problem and how you found it?
   b. Is this problem like any other problem you've solved? How?
   c. Do you think this problem could be solved in another way? What are your ideas?
   d. How did you feel while you were solving this problem? How do you feel now that you have found a solution?

Fig. 7

21

An anecdotal record, rating scale, or checklist might be used to record the findings from a structured interview. An audio or video recording can also be used to collect more detailed information for later analysis.

Give it a try!

Use the sample Interview Plan above as a guide and (1) select a person to interview, (2) select an appropriate problem, (3) devise your own interview plan, and (4) conduct a structured interview to gain insight into the person's problem-solving performance and attitudes.

What are some advantages of the structured interview?

Advantages of this technique include the following:

- It allows careful observation of performance or attitudes on a one-to-one basis.
- It allows the evaluator time to probe more deeply into an individual's problem-solving skills.
- It allows for a high degree of structure, or flexibility, depending on the wishes of the evaluator.
- It provides students an opportunity to give detailed information about what they are doing and thinking.
- It provides insight into a student's thinking processes that are not usually apparent from written work.

What are some disadvantages of a structured interview?

Disadvantages include the following:

- It takes a lot of time.
- It requires that questions be chosen carefully and asked at precisely the right time.
- It involves asking students questions while they are working on a problem, which may hinder their performance.
- It may not provide standard information that allows for comparison of individuals.

When and how should structured interviews be used?

This technique should be used when you want to probe more deeply into a student's thinking processes, problem-solving performance, or attitudes. It may be helpful as a diagnostic procedure for students having considerable difficulty. It is also useful when collecting data for research into the problem-solving process.

The following steps suggest a general procedure for using a structured interview:

1. Decide on the structure of the interview and select a means of recording the student's responses.

2. When the student arrives, take time to establish a friendly, relaxed atmosphere.

3. Present a problem to be solved and ask the student to talk as much as possible about what he or she is doing or thinking during the process.

4. Observe and listen as the student works on the problem, asking probing questions to clarify what the student is thinking or doing. Refrain from teaching or asking leading questions.

5. Carefully record evidence of whatever action or thought is germane to the goals being evaluated.

An alternative approach in step 3 for a reticent student might be for the interviewer to solve the problem and ask the student to tell what he thinks the interviewer is doing or thinking as the problem is being solved.

A structured interview is usually used to evaluate important problem-solving behaviors that cannot be assessed more easily using other techniques.

How does one structure an interview?

An evaluator structures an interview by devising beforehand a set of questions to use to probe the student's problem-solving skills in each of the following areas, as shown in the sample interview plan in figure 7.

1. Understanding the problem question and conditions
2. Selecting and using data
3. Selecting and using strategies
4. Solving and answering the problem
5. Checking the reasonableness of the solution

Each of these phases of problem solving can be explored in turn as the student proceeds to solve the problem.

TECHNIQUE 2: USING SELF-ASSESSMENT DATA FROM STUDENTS

Observation is a useful technique for evaluating a number of important problem-solving performance and attitude goals. However, there are other goals that can be fully evaluated only by collecting self-assessment data from students. The usefulness of such assessments depends, of course, on how candidly they report their feelings, beliefs, intentions, thinking patterns, and so forth.

One technique for collecting assessment data from students is the *student report*. Students are asked to write or dictate on a tape recorder a retrospective report on a problem-solving experience they have just completed. Usually the reports are made in response to a general question or a series of questions designed to remind them to comment on selected aspects of their problem-solving experience. These reports can be used to evaluate both performance and attitude.

Another useful technique is the *inventory*. An attitude inventory is a familiar type, but performance inventories can also be used to get students' assessment of their own problem-solving ability. An inventory usually consists of a list of items where the student checks an item if it applies and leaves it blank if it doesn't. If further gradations in judgment are desired, a scale can be introduced.

These techniques will now be described in greater detail.

Student Reports

What is a student report?

This technique involves a student's written or tape-recorded retrospective report on a problem-solving experience. Students are asked to think back and describe how they solved the problem.

A general direction or question, such as "Tell about your thinking as you describe how you solved the problem," is useful for helping the student get started. To give more direction to the report, the student might be asked to respond to the focus questions in figure 8.

Student Report: Focus Questions

Use the following questions to help you look back and describe your thinking as you worked toward a solution to the problem.

1. What did you do when you first saw the problem? What were your thoughts?
2. Did you use any problem-solving strategies? Which ones? How did they work out? How did you happen to find a solution?
3. Did you try an approach that didn't work and have to stop and try another approach? How did you feel about this?
4. Did you find a solution to the problem? How did you feel about this?
5. Did you check your answer in any way? Did you feel sure it was correct?
6. How did you feel, in general, about this problem-solving experience?

Fig. 8

Note that these reports can be focused to evaluate problem-solving performance, attitudes, or a combination of these. The following questions might be used to focus on attitudes:

1. Did you ever feel frustrated when solving the problem? Why?
2. Did you ever feel that you wanted to give up and not solve the problem? When?
3. Did you enjoy solving this problem? Why or why not?

4. Would you rather have worked by yourself or with others when solving this problem? Why?

Give it a try!

Try solving the following problem. Then use the focus questions (fig. 8) to write or record a retrospective report on your own experience.

> Problem: How many squares of all different sizes are there on an ordinary checkerboard?

What are some advantages of using student reports?

Advantages of this technique include the following:

- It provides information, supplied by the student, that isn't available from other techniques.
- It allows students to feel a part of the evaluation process.
- It gives students practice in expressing their ideas orally or in writing.
- It provides unique student-oriented information to compare with or supplement evaluative information gathered using other techniques.
- It does not take any of the teacher's time during the preparation stage.

What are some disadvantages of using student reports?

Disadvantages include the following:

- It allows for the possibly ill-advised use of the reports for grading purposes rather than for helping students improve problem-solving performance and attitudes.
- It takes considerable time for students to prepare the reports.
- It can be used only with students who have adequate reporting skills.
- It may produce incomplete or inaccurate information, since students may not remember to report some of the important things they do during problem solving.

When and how should student reports be used?

Student reports can be used to help evaluate several of the performance and attitude goals of problem-solving. For example, they give useful information on students' ability to monitor and evaluate their thinking during, and immediately after, solving a problem. They can give the teacher a fresh perspective on a students' attitudes and beliefs toward problem solving and their impressions of their individual role in group problem solving. Student reports often provide important information about the individual's use of strategies and thinking skills. They should be used when you wish to evaluate the goals above, and use the information to help students improve. They should not be used for grading purposes, since this might influence the frankness of the students' responses.

Before using this technique, try to develop a feeling among your students that their honest and thorough responses will give you important information that will enable you to help them become better problem solvers. They should be asked to complete their reports immediately following a problem-solving experience. Some may find it natural and exciting to use a tape recorder, and others may be more effective preparing a written report.

How can focus questions for a student report be developed?

The focus questions in figure 8 are only a sample. As an alternative, you may wish to limit the focus to only one phase of the problem-solving experience and ask questions only about that phase. First decide what aspect you want the students to report on. Then formulate questions that stimulate their recollections of experiences during this phase of solving the problem.

Inventories

What is an inventory?

An *inventory* is a list of items a student checks selectively to give a systematic self-appraisal of performance or attitudes. An attitude inventory might be a simple yes-no (or true-false) checklist, such as the one in figure 9 developed for the Mathematical Problem-Solving Project at Indiana University, or a more extensive attitude scale that asks for varying degrees of opinion.

The three categories that are assessed by the items in figure 9 are *willingness* to engage in problem-solving activities (items 2, 3, 5, 14, 15, 17), *perseverance* during the problem-solving process (items 1, 4, 8, 10, 16, 18), and *self-confidence* with respect to problem solving (items 6, 7, 9, 11, 12, 13, 19, 20). Items are worded to reveal positive or negative feelings as follows: *positive*—items 3, 5, 8–11, 13, 16, 17, 20; *negative*—items 1, 2, 4, 6, 7, 12, 14, 15, 18, 19.

Attitude inventories can be extensive instruments like the one in figure 9, in which reliability and validity measures have been established, or a simple teacher-made variety (fig. 10) for younger children. Such inventories have also been used to sample children's attitudes toward particular kinds of problems.

Inventories, or checklists for student self-assessment of problem-solving performance, can also be as simple or detailed as desired. For example, the inventory in figure 11 might be made by a teacher to help evaluate a student's choice and use of problem-solving strategies when solving a given problem.

Give it a try!

Complete the twenty-item attitude inventory in figure 9. Find your total

Attitude Inventory Items

Pretend your class has been given some math story problems to solve. Mark true or false depending on how the statement describes you. There are no right or wrong answers for this part.

_____ 1. I will put down any answer just to finish a problem.

_____ 2. It is no fun to try to solve problems.

_____ 3. I will try almost any problem.

_____ 4. When I do not get the right answer right away I give up.

_____ 5. I like to try hard problems.

_____ 6. My ideas about how to solve problems are not as good as other students' ideas.

_____ 7. I can only do problems everyone else can do.

_____ 8. I will not stop working on a problem until I get an answer.

_____ 9. I am sure I can solve most problems.

_____ 10. I will work a long time on a problem.

_____ 11. I am better than many students at solving problems.

_____ 12. I need someone to help me work on problems.

_____ 13. I can solve most hard problems.

_____ 14. There are some problems I will just not try.

_____ 15. I do not like to try problems that are hard to understand.

_____ 16. I will keep working on a problem until I get it right.

_____ 17. I like to try to solve problems.

_____ 18. I give up on problems right away.

_____ 19. Most problems are too hard for me to solve.

_____ 20. I am a good problem solver.

Fig. 9

score as follows: For each negatively worded item, assign a 0 if marked "true" and 1 if marked "false." For each positively worded item, assign a 0 if marked "false" and 1 if marked "true." When this inventory was given to 100 fifth graders, their mean score was 12.83 (based on 20 points). Their mean scores on the subcategories were as follows: willingness, 4.34 (based on 6 points); perseverance, 3.94 (based on 6 points); and self-confidence, 4.55 (based on 8 points). Calculate your mean scores and compare them with those of the fifth graders. A student score well below the average in any category might indicate that attention should be given to providing experiences and reinforcement to help the student in that area.

What are some advantages of using an inventory?

Advantages of this technique include the following:

- It allows students some input into the evaluation process.
- It requires very little of the teacher's time for collecting the evaluation data.

How I Feel about Problem Solving

Mark an x on the face that shows how you feel when . . .

    1. Your teacher gives you a story problem to solve.

    2. You think about the problem and find the right answer.

Fig. 10

Problem-solving Strategy Inventory

Think about your use of strategies when solving the problem and check the following that apply.

  1. _____ I didn't think about using strategies at all.

  2. _____ The idea of using strategies came to my mind, but I didn't think about it much more.

  3. _____ I looked at a strategy list, but didn't try a strategy.

  4. _____ I looked at a strategy list and picked a strategy, which I tried.

  5. _____ I didn't look at a list, but just thought of a strategy to try.

  6. _____ I used at least one strategy and it helped me find a solution.

  7. I tried the following strategies:

    _____ guess and check   _____ solve a simpler problem

    _____ make a table   _____ work backward

    _____ look for a pattern   _____ draw a picture

    _____ make an organized list _____ write an equation

    _____ other_____

Fig. 11

- It provides for the systematic collection of data on preselected aspects of problem-solving performance or attitude.
- It provides a personal student assessment that the teacher can use to supplement other evaluation data.

28

What are some disadvantages of using an inventory?

Disadvantages include the following:

- Its accuracy depends on the quality of a student's insight into his or her performance or attitude.
- It allows for possible misinterpretation or failure to be candid.
- It is easy to assume unwarranted validity and reliability for an inventory.

When and how do you use an inventory?

Inventories are especially useful for measuring student attitudes and beliefs related to problem solving. They can also be useful in evaluating those aspects of problem-solving performance that can be accurately assessed by the student. They also indicate the students' reaction to the teacher's particular method of teaching problem solving. Inventories should be used when student input would be particularly useful, but *they should probably be used only in conjunction with other evaluation techniques, such as teacher observations and tests.*

Be sure to have a specific purpose for administering the inventory. Also, establish a climate in which the students feel that their honest responses are important and will enable you to help them become better problem solvers. After administering the inventory, tabulate the results. Rather than using elaborate numerical analysis, you may want simply to group the students as high, middle, or low, and compare the measures of individuals with a representative measure of the group. This global interpretation can help you focus on the attainment of goals and the needs of your students.

TECHNIQUE 3: HOLISTIC SCORING

The amount of time required for evaluating progress in problem solving using observation and questioning methods usually makes it impractical to rely on them exclusively. Since students are often asked to show their solutions in writing, it is natural to consider how a student's written work on a problem can be used to help evaluate progress in problem solving.

Teachers have always evaluated students' written work. Usually this involves checking to see if a computational skill has been used properly to arrive at a correct answer. Typically the answer is checked first and if it is incorrect, the cause of the error is sought. When evaluating problem solving by analyzing a student's written work, the approach is similar. However, the emphasis now should be on the problem-solving process.

In this section we describe three methods for evaluating a student's written work on a problem. The first method, *analytic scoring,* involves the use of a scale to assign points to certain phases of the process. The second method, *focused holistic scoring,* enables you to assign a numerical score to the total solution of a problem based on criteria related to specific thinking processes. The third method, *general impression scoring,* uses a general

29

impression, based on the evaluator's implicit criteria, to rate a total solution numerically. The success of each method is enhanced by finding ways to ensure that students record as much as possible of the results of their thinking. Since any analysis of a student's written work that is independent of observing the student solve the problem requires some inference, all techniques for evaluating written work should be supplemented with observation and questioning or the collection of assessment data from students.

Analytic Scoring

What is analytic scoring?

Analytic scoring is an evaluation method that assigns point values (scores) to each of several phases of the problem-solving process. Thus, the first step in developing an analytic scale is to identify those phases of the problem-solving process that are of interest to you. The second is to specify a range of possible scores for each phase. The range of points that we suggest is 0–2.

An example of an analytic scoring scale appears in figure 12. This scale was developed for three phases or categories of problem solving: *understanding the problem, planning a solution,* and *getting an answer.* For each of these problem-solving categories, 0, 1, or 2 points would be assigned. Criteria are given for awarding points in each category.

Analytic Scoring Scale	
Understanding the Problem	0: Complete misunderstanding of the problem 1: Part of the problem misunderstood or misinterpreted 2: Complete understanding of the problem
Planning a Solution	0: No attempt, or totally inappropriate plan 1: Partially correct plan based on part of the problem being interpreted correctly 2: Plan could have led to a correct solution if implemented properly
Getting an Answer	0: No answer, or wrong answer based on an inappropriate plan 1: Copying error; computational error; partial answer for a problem with multiple answers 2: Correct answer and correct label for the answer

Fig. 12

Give it a try!

Try your hand at using this analytic scale! Study the problem below and its

Problem: Tom and Sue visited their grandparents' farm. In the barnyard there were some chickens and pigs. Tom said: "I see 18 animals in all." Sue answered: "Yes, and altogether they have 52 legs." How many chickens and how many pigs were in the barnyard?

One Possible Solution:

Make a Table.

Chickens	Pigs	Total Legs
18	0	36 (much too low)
12	6	48 (still too low)
10	8	52 (That's it!!)

solution. Then use the analytic scale (fig. 12) for the three students' papers (figs. 13–15). Decide how many points you would give each student for each of the three categories (U-Understanding; P-Planning; A-Answer).

There should be no doubt that Anne deserves a better score than Mike and Rosita. But how much better? And how do Mike and Rosita compare? Let's consider their work individually, looking at Anne's work first.

Anne's work. It is evident from what Anne wrote that she has identified all the important data: namely, that there are 18 animals and 52 legs. So, Anne gets *2 points for understanding*. It appears that her plan was to use a guess-and-check strategy. Her initial guess seems to have been 9 chickens and 9 pigs (9c 9p). She then correctly checked this guess, finding that it resulted in 54 legs instead of 52. She should get *2 points for her plan*. Using this information, Anne adjusted her first guess by exchanging one pig for one chicken, yielding 10 chickens and 8 pigs—the correct answer. Finally, Anne labeled her answer correctly; thus, *2 points for her answer*. Her final rating: *U - 2, P - 2, A - 2* (an overall score of 6—the maximum).

Rosita's work. Rosita's work has merit, even though she obtained a wrong answer. Like Anne, she did use all the important information; let's give her *2 points for understanding*. Scoring her work becomes more difficult at this point. It could be argued that she deserves *1 point for her plan* (the plan has some merit) and 0 points for her answer, since it is not correct. However, we would give Rosita *1 point for her answer* because even though it is wrong, it may be the result of a copying error. Thus, one evaluation of her work would be *U - 2, P - 1, A - 1* (overall score, 4).

31

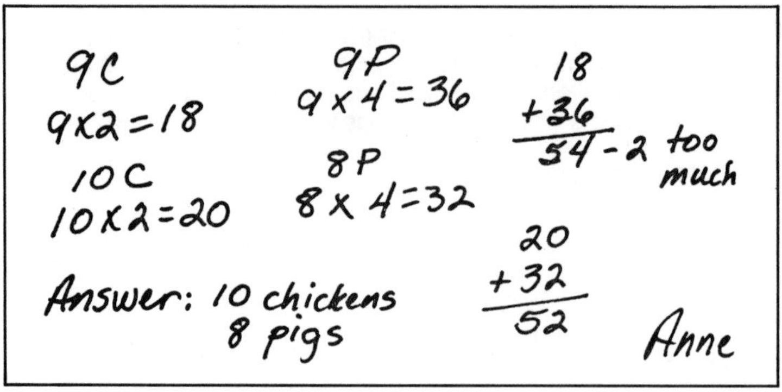

Fig. 13

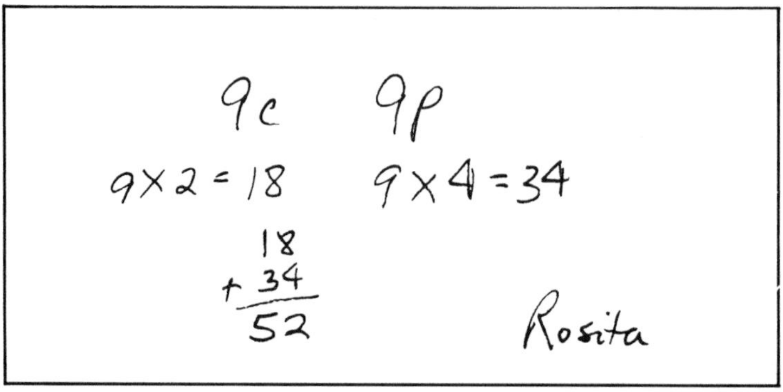

Fig. 14

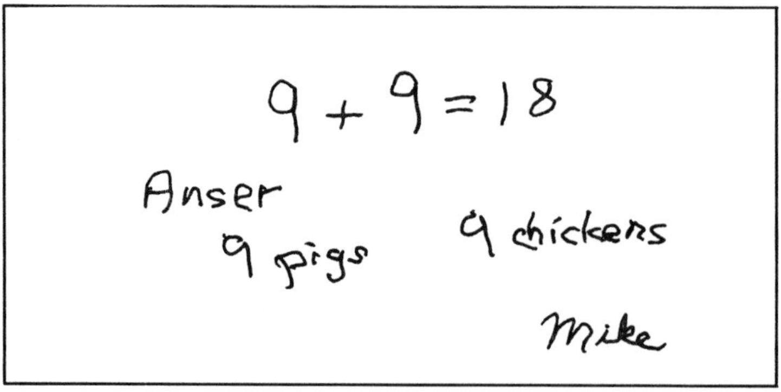

Fig. 15

Mike's work. Mike's work indicates that he ignored part of the relevant information in the problem, namely that there were 52 legs. We would give him *1 point for understanding.* His plan seems to have been to find two numbers whose sum is 18. This plan could lead somewhere (it did for Anne), but it is only partially correct. He did not check his numbers against all the

32

information. Mike gets *1 point for his plan*. Mike's answer is labeled correctly even though the answer itself is incorrect. That is, his answer has some merit although it is wrong. Give him *1 point for his answer*. His rating, then, is *U - 1, P - 1, A - 1* (overall score, 3).

What are the advantages of analytic scales?

Among the advantages of analytic scales, five are especially prominent:

- They consider several phases of the problem-solving process, not just the answer.
- They provide a means for assigning numerical values to students' work.
- They help the teacher pinpoint specific areas of strength and weakness.
- They provide specific information about the effectiveness of various instructional activities.
- They allow for the differential weighting of categories that make up the scale.

What are the disadvantages of analytic scales?

Disadvantages include the following:

- In some cases an individual student's written work might not provide enough information about his or her thinking processes to enable the teacher to assign points to one or more categories with confidence.
- The categories making up an analytic scale must receive direct attention during instruction. Thus, the scale must be carefully correlated with the instructional program.
- Comparisons of students' scores must be made with great care. That is, two students can have the same overall scores even though they have performed quite differently in solving a problem. For example, a student scoring 2-1-1 has performed very differently from a student scoring 2-2-0.

When should an analytic scale be used?

The analytic scoring method is based on the belief that problem-solving evaluation should involve more than simply checking the final answer. An analytic scale enables you to assess students' performance with respect to key *predetermined* phases of the problem-solving process. As a result, it becomes possible to identify specific areas of strength and weakness and to assess the effectiveness of specific instructional activities. With these considerations in mind, analytic scales are most useful under the following conditions:

- When it is desirable to give students feedback about their performance in key categories associated with problem solving
- When it would be useful to have diagnostic information about their specific strengths and weaknesses

- When you are interested in identifying specific aspects of problem solving that may require additional instructional time
- When you have enough time to carefully analyze each student's written work

How do you develop an analytic scale?

The following steps should be taken to develop an analytic scale:

1. Identify the phases (categories) of the problem-solving process that you want to be the focal points of your evaluation.
2. Decide on the number of points each category is worth.
3. Determine what criteria must be met for a student's work to earn points in a category.

The 6-point scale described here is but one of several possibilities. Another possibility would be to assign points to each (or to several) of the seven thinking skills discussed in an earlier section of this booklet. Still another would be to develop a scale around the "model" for problem solving described in your classroom mathematics text. Whatever scale you decide to use or develop, strive for as much consistency in its use as possible. That is, it is more important to use the scale in a consistent manner than to be overly concerned with the distinction between, say, *U - 2, P - 1, A - 1* and *U - 2, P - 1, A - 0*. As you gain experience in using it, you will develop a better feel for its strengths and limitations.

Another suggestion for constructing an analytic scale is to consider giving more weight to some categories than others. For example, if you use the analytic scale described earlier and you are particularly interested in how well your students are able to identify and use the necessary data in problems, you may decide to give the *understanding* category twice the weight of the other two categories. In other words, if you have stressed a particular phase of the problem-solving process during instruction, you should consider stressing that phase during evaluation as well.

Finally, it seems appropriate to mention again that analytic scales should probably be used in conjunction with other evaluation procedures, especially informal observations and questioning.

Focused Holistic Scoring

What is focused holistic scoring?

Focused holistic scoring produces one score for a student's work on a problem. It is *holistic* because it focuses on the total solution, not just on the answer. It is *focused* because one number is assigned according to specific criteria related to the thinking processes involved in solving problems. Unlike the analytic scale described above, focused holistic scoring does not involve assigning points to each of several categories of thinking processes but rather assigns a single score for the entire solution. An example of a focused holistic scoring scale appears in figure 16.

34

Focused Holistic Scoring Point Scale

0 Points

These papers have one of the following characteristics:

- They are blank.
- The data in the problem may be simply recopied, but nothing is done with the data or there is work but no apparent understanding of the problem.
- There is an incorrect answer and no other work is shown.

1 Point

These papers have one of the following characteristics:

- There is a start toward finding the solution beyond just copying data that reflects some understanding, but the approach used would not have led to a correct solution.
- An inappropriate strategy is started but not carried out, and there is no evidence that the student turned to another strategy. It appears that the student tried one approach that did not work and then gave up.
- The student tried to reach a subgoal but never did.

2 Points

These papers have one of the following characteristics:

- The student used an inappropriate strategy and got an incorrect answer, but the work showed some understanding of the problem.
- An appropriate strategy was used, but—
  - a) it was not carried out far enough to reach a solution (e.g., there were only 2 entries in an organized list);
  - b) it was implemented incorrectly and thus led to no answer or an incorrect answer.
- The student successfully reached a subgoal, but went no further.
- The correct answer is shown, but—
  - a) the work is not understandable;
  - b) no work is shown.

3 Points

These papers have one of the following characteristics:

- The student has implemented a solution strategy that could have led to the correct solution, but he or she misunderstood part of the problem or ignored a condition in the problem.
- Appropriate solution strategies were properly applied, but—
  - a) the student answered the problem incorrectly for no apparent reason;
  - b) the correct numerical part of the answer was given and the answer was not labeled or was labeled incorrectly;
  - c) no answer is given.
- The correct answer is given, and there is some evidence that appropriate solution strategies were selected. However, the implementation of the strategies is not completely clear.

4 Points

These papers have one of the following characteristics:

- The student made an error in carrying out an appropriate solution strategy. However, this error does not reflect misunderstanding of either the problem or how to implement the strategy, but rather it seems to be a copying or computational error.
- Appropriate strategies were selected and implemented. The correct answer was given in terms of the data in the problem.

Fig. 16

This scoring scale is similar to the analytic scale in that descriptors of the problem-solving process are used as guidelines for assigning points. The specific criteria given in the focused holistic scale were developed over a period of several months by teachers scoring students' written work. Different criteria might be used, or the ones above could be rearranged to reflect different emphases. For example, the scale above shows 4 points for papers with a wrong answer due to a computational error. Some teachers prefer to assign 4 points only to those papers where the total process and answer are correct. The criteria can certainly be modified to meet your individual needs or preferences.

Give it a try!

A problem together with one solution appears in figure 17. Study both, and then use the focused holistic scale (fig. 16) to assign points to the two students' papers (figs. 18 and 19).

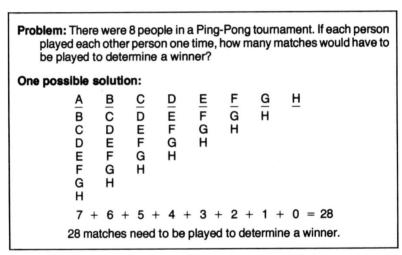

Fig. 17

It's obvious in looking at these papers that Carmen has an incorrect answer and Jack's is correct. However, we would assign a 3 to Carmen's paper and a 2 to Jack's using the criteria above. Carmen apparently misunderstood or ignored the condition that each person played each other person once. However, her strategy of drawing a picture could have led to the correct solution. For Carmen's paper, the first criterion in category 3 applies. That is, she implemented a solution strategy that could have led to the correct solution, but she misunderstood or ignored a condition. For Jack's paper, the multiplication fact $8 \times 2 = 16$ suggests there were 8 games, not 8 people. Also, repeatedly dividing by 2 seems to suggest that he understood half of the people would lose each time, but the reason he stopped at 4 is not clear. We would assign a 2 to Jack's paper using the criterion in category 2 that the correct answer is given but the solution method is not understandable.

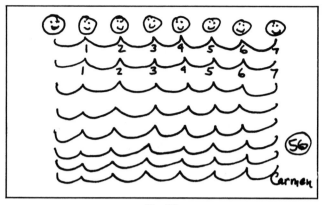

Fig. 18

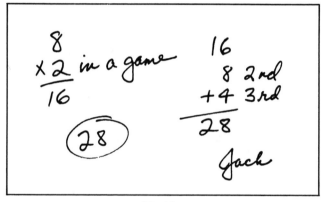

Fig. 19

What are the advantages of focused holistic scoring scales?

Focused holistic scoring scales have these advantages:

- They allow for a relatively fast assessment of a student's written work.
- They focus on the processes used, not just the answer.
- They provide specific criteria to guide the scoring of papers.
- They provide one score to describe performance.

What are the disadvantages of focused holistic scoring scales?

Disadvantages include the following:

- They do not permit the pinpointing of specific strengths and weaknesses.
- Some students' written work may not provide enough information about their thinking processes to enable the teacher to assign points with confidence.
- They do not permit a differential weighting of the thinking processes involved in solving problems.

37

When should a focused holistic scoring scale be used?

Focused holistic scoring is most appropriate when you are interested in a general rating of the processes used, and explicit criteria are needed or wanted to guide the assigning of points. Focused holistic scales are most appropriate as a precursor or as a follow-up to other evaluation techniques aimed at identifying students' strengths and weaknesses relative to the thinking processes involved. Such a scale might be used in assessing performance on chapter or semester examinations where a large number of papers need to be evaluated and the evaluation emphasis is on the total problem-solving process. The existence of specific criteria for assigning points promotes consistency in evaluating written performance. Therefore, focused holistic scales are also appropriate for large-scale evaluations (for example, district-wide assessments) where a number of scorers will be used and the reliability of the scoring procedure is important. In large-scale evaluations, "anchor papers" should be identified for each point category. Anchor papers exemplify the criteria for a point category. For example, the sample correct solution in figure 17 could serve as an "anchor" for the four-point category. (See the references for sources of information on using holistic scoring schemes in large-scale evaluations.)

How do you develop a focused holistic scoring scale?

The first step is to decide what point range will be used. Since specific criteria need to be developed for each point, five or fewer categories seem most appropriate. When more than five categories are used, it is difficult to develop distinct criteria, and the reliability of assigning points often drops. The second and most crucial phase is developing the criteria for each category. In a manner similar to developing analytic scales, you should identify behaviors associated with the thinking processes involved and then assign these behaviors to a point category. If you develop your own scale, try it with several samples of students' work before using it with all papers. When new scales are developed, some revisions are often needed and this becomes apparent only when you try using the scale.

General Impression Scoring

What is general impression scoring?

General impression scoring is a technique in which the evaluator studies a student's solution to a problem and relies on a general impression to rate it on a scale, such as 0 to 4. This is the least complicated of all holistic scoring methods, since unlike the methods described earlier, no written criteria or rating sheets are prepared or used. However, the evaluator uses implicit criteria, based on a subjective view of the important components of a problem solution and on experience gained from looking at a variety of problem solutions. Because of this, *general impression scoring should not be used by teachers who have had limited experience assessing problem solving.*

When using this method, you could also write comments or questions on students' papers about particular aspects of the solution.

Give it a try!

Study the problem and solution given below. Then assign a general impression rating on a 0 to 4 scale to each student paper (figs. 20 and 21).

Problem: Sandy made 3 free throws out of every 5 shots during the basketball season. How many free throws would you expect her to make in 30 shots?

One possible solution:

Free Throws	3	6	9	12	15	18
Shots	5	10	15	20	25	30

Sandy would be expected to make 18 free throws.

As we look at Teri's paper, we see that she began with 3 of 5 and listed equal ratios. It appears that after she wrote 9 of 15, she doubled the numbers and jumped to 18 of 30. She has given the correct answer in a complete sentence. Our general impression is that she understood the problem, used an effective strategy, and has arrived at the correct answer. This suggests a rating of 4.

Jim's answer is incorrect, but as we study his written work we observe that finding the number of groups of five shots in thirty would tell us the number of groups of three shots that Sandy would be expected to make. We see that Jim started to multiply 5 by 3, but marked it out. It appears that he might have had a reasonable idea about how to solve the problem, but became confused as he tried to carry it out. Our general impression is that his solution merits a rating of 1 or 2. We choose 2.

Name ___Teri___   Grade __5__

3   5
6   10   She would make
9   15   18 free throws.
18   30

General Impression Rating (0–4) _____

Fig. 20

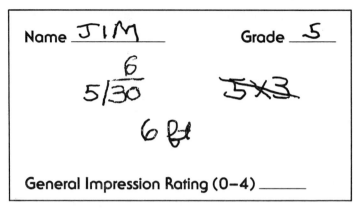

Name __JIM__   Grade __5__

$$5\overline{)\smash{\overset{6}{}}30}$$

5×3

$6 \; \text{R4}$

General Impression Rating (0-4) _____

Fig. 21

What are some advantages of general impression scoring?

Advantages of this method include the following:

- It requires no advanced preparation of criteria, checklists, or scoring sheets.
- It allows an intuitive consideration of the details of the solution in addition to a correct or incorrect answer.
- It allows for quick scoring of a solution to a problem, without detailed analysis.
- It is a method familiar to students, and one they understand.
- It may be as reliable for rating purposes as other, more complicated methods.

What are some disadvantages of general impression scoring?

Disadvantages include the following:

- It involves using implicit criteria that may not be based on a careful analysis of the problem-solving process.
- It allows for a scoring decision that may be based only on certain aspects of the solution.
- It does not provide a systematic means of giving students feedback on aspects of problem solving in which they may need improvement.
- It may encourage reliance on a number rating rather than on a careful analysis of a student's solution.
- It requires that the user have considerable experience in assessing problem solving.

When and how might general impression scoring be used?

General impression scoring is useful when you want to give students feedback on a problem solution and there isn't much time for evaluation. It

40

can be used selectively in the classroom for short assignments and for chapter or unit tests. It can also be used for rating or ranking a large number of student papers, such as in a standardized testing or contest situation. General impression scoring is efficient in situations where determining a numerical rating or grade is the goal and obtaining information needed to help improve students' problem-solving skills is less important.

Before using this method, discuss it with students and explain its purpose. They are accustomed to general impression ratings of movies (the four stars system) and other ratings on a scale of 0 to 10, so this method is usually easy for them to understand.

Try to avoid the pitfall of focusing too narrowly on only certain parts of the student's solution. Be on the lookout for, and accept the use of, strategies and resulting solutions other than the ones you are expecting. Also note that although general impression scoring does not require a teacher to make written comments on a student's paper, the teacher should, whenever possible, give feedback in addition to a single numerical evaluation.

TECHNIQUE 4: MULTIPLE-CHOICE AND COMPLETION TESTS

The written test is probably the most frequent means of evaluating problem-solving performance. Because we rely so much on written tests, it is important that they give as much quality information as possible about a student's ability to solve problems. In general, there are two important questions a written test can help answer.

- Can a student successfully carry out the various thinking processes involved in solving a problem?
- Can the student find the correct answer to a given problem?

One purpose of an assessment program for problem solving should be to obtain information about students who do not perform well at getting correct answers. Test items that provide answers to the first question above can give teachers data that will help them determine what instructional experiences are needed to improve a student's ability to get correct answers.

Since most tests are made up of items that evaluate a student's ability to get the correct answer, the next two sections will focus instead on the development and use of items that assess the thinking processes involved in solving problems. Two types of written test items that are particularly useful in evaluating a student's thinking processes are *multiple-choice* and *completion*. To prepare such items, clear statements describing the thinking processes and statements of objectives for each process are needed. Figure 22 reviews the thinking processes described earlier and includes some possible objectives. The list of objectives is not exhaustive but includes those that can be measured reliably and that seem to address the thinking process with which they are associated. The sample multiple-choice and completion test

41

Problem-solving Thinking Processes and Objectives

Thinking Processes	Sample Objectives
1. Understand/formulate the question in the problem.	Given a problem, select, write, or state in your own words the question that will be answered when a solution is found.
2. Understand the conditions and variables in the problem	Select/identify the key conditions and variables useful in understanding and solving the problem.
3. Select/find data needed to solve the problem.	a) Given a problem with unneeded data, identify the data needed to find a solution. b) Given a problem with missing data, find data needed for solving the problem.
4. Formulate subproblems and select an appropriate solution strategy to pursue.	a) Given a multiple-step or process problem, formulate/select subproblems that could be solved to find the solution. b) Given a problem, select a strategy that could be used to solve the problem.
5. Correctly implement the solution strategy and attain the subgoals.	Given a story problem— a) select/draw a picture that could be used to help solve it; b) write a number sentence that could be used to solve it.
6. Give an answer in terms of the data in the problem.	Given the numerical part of the answer to a problem, write the answer in a complete sentence.
7. Evaluate the reasonableness of the answer.	Given a problem and its answer, estimate to decide if the answer is reasonable.

Fig. 22

items presented and discussed here are designed to measure the objectives for the seven problem-solving thinking skills discussed earlier.

Multiple-Choice Tests

What are multiple-choice tests?

A *multiple-choice test* is made up of items that consist of a problem or question and a list of possible solutions or answers. After reading or hearing the problem, the student selects the correct or best answer. The other possible answers, usually called distractors, often reflect common mistakes or misinterpretations, and are designed to entice students who are unsure about the correct response.

42

Multiple-choice items are versatile and can measure the ability to get a correct answer as well as the ability to use problem-solving thinking skills. In this section, we will focus on problem-solving thinking skills. Consider the following example.

Problem-solving thinking process. Understand the conditions and variables of the problem.

Objective. Select/identify the key conditions and variables useful in understanding and solving the problem.

Multiple-choice item

Which statement best describes the meaning of the underlined phrase in this problem?

Problem: Michelle wants to buy six of her favorite tapes and needs to decide how much money to take to the record store. The tapes are on a special sale in which the first tape costs $6.20 and <u>each successive tape purchased costs $0.15 less than the previous one.</u> How much will the 6 tapes cost?

    A. Each tape costs $6.20.

    B. Each tape after the first costs $0.15.

    *C. Each tape costs $0.15 less than the one bought before it.

    D. Each tape costs $0.15 less than the one bought after it.

Note: * indicates the correct response.

Examples of multiple-choice items that are designed to measure the other problem-solving thinking skills will be given later in this section.

Give it a try!

Select a phrase from the problem below and write a multiple-choice item like the one above that measures a student's ability to understand the conditions and variables in the problem.

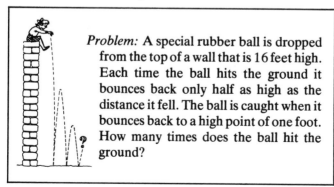

Problem: A special rubber ball is dropped from the top of a wall that is 16 feet high. Each time the ball hits the ground it bounces back only half as high as the distance it fell. The ball is caught when it bounces back to a high point of one foot. How many times does the ball hit the ground?

What are some advantages of using a multiple-choice test?

Advantages of this technique include the following:

- A wide variety of abilities can be measured.
- Items can be specific and easily understood.
- Student difficulties can be diagnosed by analyzing incorrect responses.
- Scoring and interpretation are easy.

What are some of the disadvantages of using a multiple-choice test?

Some disadvantages of this technique are as follows:

- It is difficult to write good items that measure higher-level thinking processes.
- It is difficult to find good incorrect responses for some items.
- It takes a lot of time to construct a good multiple-choice test.
- The test is not useful for measuring the ability to organize and present ideas.
- The format makes it easy for some students to guess rather than to think through the situation.

When and how should multiple-choice tests be used?

Multiple-choice tests are better suited for measuring problem-solving performance than for measuring attitudes. They can be used to measure problem-solving thinking processes and the ability to find the correct answer. Multiple-choice tests can be used in a classroom setting to evaluate the results of instruction in problem solving. They are particularly useful in large-scale evaluation projects where both correct answers and thinking skills are to be evaluated.

You may wish to add an interpretive dimension to multiple-choice tests by allowing students to write qualifying remarks about individual items. For example, those who are unsure about the meaning of any part of an item could explain their interpretation and tell why they chose their answer. These explanations can help you understand how a student is thinking about the aspects of problem solving involved in the item.

Multiple-choice items should not be the exclusive means for evaluating problem solving. If you wish to analyze carefully a student's procedure for solving a given problem, to gain specific insight into a student's ability to use higher-level problem-solving thinking skills, or to assess ability to organize and present ideas about a solution, other evaluation techniques should also be used.

How do you develop multiple-choice items to measure problem-solving thinking skills?

Before you prepare a multiple-choice item, specify exactly what ability you want the item to measure and prepare it so that it does, indeed, evaluate this ability. You can use the following procedure:

1. Identify the outcomes you want to measure.

2. Prepare specific objectives for these outcomes.

3. Write items to measure outcomes in terms of the specified objectives, carefully selecting distractors that will provide diagnostic information.

It is important to judge carefully whether an item is valid, that is, whether it measures what it was designed to measure. Analyzing exactly what the item requires the student to do or know is a good approach to checking validity. Valuable information is gained by asking students and other teachers to participate in this analysis.

Sample multiple-choice items

To illustrate how multiple-choice items can be used to evaluate a student's ability to use the seven problem-solving thinking skills, we have included twenty-one sample assessment items together with guidelines for writing additional items. Three items are given for each of the seven thinking processes: one item uses a one-step problem, one uses a multiple-step problem, and one uses a process problem. These types of problems are described in the chapter "What Are You Trying to Evaluate?" under Goal 7. Three important problem characteristics, discussed earlier, are systematically varied in these test items:

1. The type of problem and the strategy used to solve it

2. The content or the types of numbers used in the problem

3. The source from which data need be obtained for solving the problem

Problem-solving thinking skill 1. Understand the question in the problem.

Objective. Given a problem, select, write, or state in your own words the question that would be answered when a solution is found.

Multiple-choice items 1

1.1 One-step problem item

Which statement, A, B, C, or D, is another way of asking what you are trying to find out in this problem?

Problem: Jack and Denise divided the construction paper evenly among the 24 children in the room. Altogether they gave out 144 pieces of paper. How many pieces of paper did each child receive?

A. How many pieces of construction paper did Jack and Denise give out altogether?

*B. How many pieces of construction paper was each of the 24 children given?

C. How many children received the same number of pieces of construction paper?

D. How many pieces of construction paper did Jack and Denise receive altogether?

1.2 Multiple-step problem item

Which statement, A, B, C, or D, is another way of asking what you are trying to find out in this problem?

Problem: Suppose you pay $1.25 per pound for an unprocessed side of beef weighing 500 pounds. The butcher processes it and removes the waste, which is 33% of the total weight. He packages and freezes it for you at no extra charge. How much are you actually paying for a pound of processed beef?

A. What is 33% of 500 pounds?

B. How many pounds of waste are in a 500-pound side of beef?

*C. How much does one pound of the edible meat cost if you know the total cost for all of the meat and the proportion of waste?

D. How much does it cost to buy 500 pounds of beef?

1.3 Process problem item

Which statement is another way of asking what you are trying to find out in this problem?

Problem: Jack and Jill collect old buckets. Jill has 3 more buckets than Jack. Together they have 21 buckets. How many buckets has Jack collected?

*A. Altogether, how many buckets does Jack have?

B. How many fewer buckets has Jack collected?

C. How many more buckets has Jill collected than Jack?

D. How many more buckets does Jack need to collect to have the same number as Jill?

Possible guidelines for assessment items—thinking skill 1

1. Always position the question at the end of the story situation.

2. Use situations where all data needed to solve the problem are given in the story rather than, for example, in a table or chart.

Problem-solving thinking skill 2. Understand the conditions and variables in the problem.

Objective. Interpret the meaning of key information given in the problem.

Multiple-choice items 2

2.1 One-step problem item

Which statement best describes the meaning of the underlined phrase in this problem?

Problem: Carrie and her three girl friends collected aluminum cans for 6 months. At the end of that time they took the cans to a recycling center and received a total of $132, which they divided among themselves. How much money did each girl receive?

A. Each person received $132.

B. Each person got the same amount of money after it was divided.

C. They want to earn a total of $132.

*D. The 4 girls received $132 for all of the cans.

2.2 Multiple-step problem item

Which statement best describes the meaning of the underlined phrase in this problem?

Problem: Carrie is allowed to watch television for 35 hours each week. If she watches for 20 hours on the weekend, how many hours, on the average, can she watch television each weekday?

A. She watches 10 hours on Saturday and 10 hours on Sunday.

B. She watches a total of 20 hours on Saturday and 20 hours on Sunday.

C. She watches, at most, 20 hours on the weekend.

*D. She watches a total of 20 hours on Saturday and Sunday.

2.3 Process problem item

Which statement best describes the meaning of the underlined phrase in this problem?

Problem: Uncle Fred asked Patty how many chickens and pigs she had on her farm. She said she had 18 in all, and "if you count all their legs, you get 58." Uncle Fred said, "I know how many of each there are." Can you tell how many of each there are?

A. The animals have a total of 58 legs.

B. There are 58 animals on the farm.

C. There are more legs than animals.

*D. The legs on all of the chickens and pigs were counted.

Possible guideline for assessment items–thinking skill 2. The condition to be interpreted should be underlined in the problem statement.

Problem-solving thinking skill 3. Select or find the data needed to solve the problem.

Objective. Given a problem with or without missing or unneeded data, identify or find the data needed for solving the problem.

Multiple-choice items 3

3.1 One-step problem item

Which data will you use to solve this problem?

Problem: Ned lives in Hooterville. His brother, Ted, lives in Scooterville. When Ned visits Ted, he always drives through Dooderville. How far does Ned drive, one way, to see Ted?

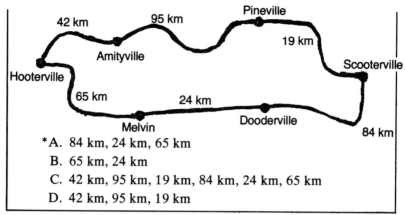

*A. 84 km, 24 km, 65 km
 B. 65 km, 24 km
 C. 42 km, 95 km, 19 km, 84 km, 24 km, 65 km
 D. 42 km, 95 km, 19 km

3.2 Multiple-step problem item

Which data do you need to solve this problem?

Problem: Mr. and Mrs. Baker and their three children bought tickets for a play. Adult tickets cost $3 each and children's tickets cost $2 each. How much did the Bakers pay for tickets altogether?

   A. All you need are the prices of the tickets.
   B. The only data you need are the number of people who bought tickets.
   C. The only data you need are the total prices for the tickets and the number of adults.
  *D. The only data you need are the number of adults, the number of children, and the price of adult and children's tickets.

3.3 Process problem item

Which data will you use to solve this problem?

Problem: Wilma's mother told her she could order 1 sandwich and 1 drink for her lunch. How many different lunches could Wilma order?

MENU		
Sandwiches	*Drinks*	*Desserts*
cheese	milk	peaches
hot dog	tea	frozen yogurt
peanut butter	apple juice	granola bar
	orange juice	cookie

49

A. names for the 3 sandwiches and 4 desserts
B. sandwiches, drinks, desserts
*C. names for the 3 sandwiches and 4 drinks
D. cheese sandwich and milk

Possible guideline for assessment items—thinking skill 3. Students should be required to select data from different sources.

Problem-solving thinking skill 4. Formulate subproblems and select an appropriate solution strategy to pursue.

Objective. Given a problem, select a solution strategy that could be used to help solve the problem.

Multiple-choice items 4

4.1 One-step problem item

Which is an appropriate method for solving the following problem?

Problem: Chad is 10 years old. He bought baseball tickets for himself and his younger brothers, Dan and Stan. How much did he pay altogether for the tickets?

Baseball Tickets	
Adults	$6.50 each
Children under 12	$3.25 each

A. Draw a picture.   *C. Use multiplication.
B. Use subtraction.   D. Use division.

Note: This item asks the student to identify an appropriate solution strategy.

4.2 Multiple-step problem item

Which is an appropriate *first step* in solving the following problem?

Problem: Box seats cost $8 each and grandstand seats cost $5 each. Diane ordered 3 box seats and 6 grandstand seats. What was her total cost for the tickets?

A. Find the total number of seats.
*B. Find the total cost for the grandstand seats and the total cost for the box seats.
C. Find the total cost for the tickets.
D. Find the total number of tickets and the total cost of the tickets.

Note: This item asks the student to identify an appropriate subproblem.

50

4.3 Process problem item

Which of the following solution attempts is most likely to lead to a correct solution for the following problem?

Problem: Jerry was mowing his lawn when he noticed that Kara was also mowing her lawn. They stopped to talk and they learned that Jerry mows his lawn every 8 days and Kara mows her lawn every 6 days. In how many days will they next be mowing their lawns together again?

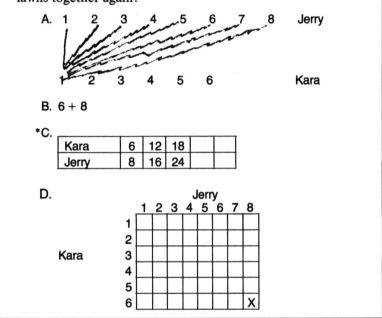

B. 6 + 8

*C.

Kara	6	12	18		
Jerry	8	16	24		

Possible guidelines for assessment items—thinking skill 4

1. Assessment items for multiple-step problems should focus on whether students can identify an appropriate first step toward finding the solution.
2. Partially completed solution strategies should be used for the distractors for process problems rather than strategy names since strategy names may not be universally accepted (e.g., *make a table* could be interchanged with *make a chart*).

Problem-solving thinking skill 5. Correctly implement the solution strategy and solve subproblems.

Objective. Given a problem, carry out a solution strategy to help find a solution.

Multiple-choice items 5

5.1 One-step problem item

Use the solution method given, and find the answer to this problem.

Problem: Ping-Pong balls come in packages of two different sizes. The small package holds 3 balls and the large one holds 4 balls. Tracey saved her money and bought 5 large packages of balls. How many Ping-Pong balls did she buy?

Solution Method: Use multiplication. ＿＿ × ＿＿ = ＿＿

    *A. 20 balls   C. 7 balls
     B. 12 balls   D. 15 balls

5.2 Multiple-step problem item

Which solution method is correct for the following problem?

Problem: Liza bought all of the available trim fabric and 1 1/5 yd. of lace. She paid for the items with two $1 bills. How much change should she get?

	Remnant Sale	
trim	$4\frac{1}{2}$ yd.	20¢ per yd.
lace	$2\frac{3}{4}$ yd.	35¢ per yd.
ribbon	$6\frac{1}{5}$ yd.	15¢ per yd.

A. $20 + 35 + 55 \rightarrow 2.00 - 0.55 = 1.4$

B. $4\frac{1}{2} + 1\frac{1}{5} = 4\frac{5}{10} + 1\frac{2}{10} = 5\frac{7}{10}$

*C. $4\frac{1}{2} \times 20 = \frac{9}{2} \times 20 = 90$ $\Big\}$ $\rightarrow 90 + 42 = 132$

     $1\frac{1}{5} \times 35 = \frac{6}{5} \times 35 = 42$ $\rightarrow 2.00 - 1.32 = 0.68$

D. $4\frac{1}{2} + 1\frac{1}{5} = 4\frac{5}{10} + 1\frac{2}{10} = 5\frac{7}{10}$ $\Big\} \rightarrow 55 \times 5\frac{7}{10}$

           $30 + 25 = 55$

                 $= 55 \times \frac{57}{10}$

                 $= 5.5 \times 57 = 313.5$

5.3 Process problem item

Complete the table below to find the answer for this problem.

Problem: The town of Rumorsville (pop. 21 845) was known for the speed with which a story could spread through the town. Each person who heard a rumor would tell it  to 4 other people in one hour and then tell it to no one else. One morning the town clerk heard a story. How long did it take for everyone in town to hear the story?

Hours elapsed								
Number of new people told the rumor								
Total number who have heard the rumor								

A. 21 384 people   C. 8 hours

*B. 7 hours   D. 16 384 people

Possible guidelines for assessment items—thinking skill 5. We recommend two methods for assessing this thinking skill. The first requires students to complete a partially completed solution strategy and choose the correct answer. Items 5.1 and 5.3 illustrate this. The second method requires students to select the correct completed solution strategy. Item 5.2 illustrates this.

Problem-solving thinking skill 6. Give an answer in terms of the data given in the problem.

Objective. Given a problem, express the answer correctly in terms of the question and data in the problem.

Multiple-choice items 6

6.1 One-step problem item

Which is the correct answer for this problem?

Problem: Kara worked on 2 Saturdays mowing lawns and washing cars. She washed 5 cars the first Saturday and 4 the next. Each  Saturday she mowed 6 lawns. Altogether, how many lawns did she mow and cars did she wash on the 2 Saturdays?

A. $21   C. 21 cars

B. 21 lawns   *D. 21 lawns and cars

6.2 Multiple-step problem item

Which is the correct answer for this problem?

Problem: An average oil well produces 16.8 barrels each day. How many gallons are produced in 1 year if there are 42 gallons in one barrel?

    A. 257 544 barrels   *C. 257 544 gallons

    B. $257 544   D. 257 544 days

6.3 Process problem item

Which is the correct answer for this problem?

Problem: A college basketball tournament has 52 teams participating. If a team loses a game, it is out of the tournament. How many games need to be played to determine a champion?

    A. 51 losses   C. 51 players

    B. 51 teams   *D. 51 games

Possible guideline for assessment items—thinking skill 6. Items 6.1, 6.2, and 6.3 illustrate the form we recommend for the distractors. All distractors should have the same number but different units or referents.

Problem-solving thinking skill 7. Evaluate the reasonableness of the answer.

Objective. Given a problem and its answer, estimate or otherwise decide if the answer is reasonable.

Multiple-choice items 7

7.1 One-step problem item

Which statement describes best why the answer given for this problem is *not* reasonable?

Problem: Carlos saved money from his paper route for 7 months to buy a new bicycle. After 7 months he had saved $125. The total cost of the bicycle, including tax, was $94.89. How much money did Carlos have after he bought the bike?

Answer: $219.89

A. The bike cost $94.89.

B. If he had $125, he spent $94.89.

*C. If he had $125 and spent $94.89, the amount left must be less than $125.

D. If the bike cost $94.89, then he saved more money than the actual cost of the bicycle.

7.2 Multiple-step problem item

Which statement best describes why the answer given for this problem is reasonable?

Problem: Renee started a special savings account. She put $78 in the account for one year. At the end of the year she had earned 10 1/2% interest. How much money did she have in her savings account at the end of the year?

Answer: $86.19

A. The answer is greater than $78.

B. 86 is almost 10 greater than 78.

C. $86.19 rounds to $86; 86 − 78 = 8; to the nearest 10, 8 rounds to 10.

*D. 10% of $80 is $8; 78 + 8 = 86.

7.3 Process problem item

Which statement best describes why the answer given for this problem is *not* reasonable?

Problem: Steve, Mike, and Holly took turns driving home from camp. Holly drove 80 km more than Mike. Mike drove 3 times as far as Steve. Steve drove 50 km. How long was the total drive?

Answer: 130 km

*A. Mike drove 150 km himself.

B. Holly drove 80 km more than Mike.

C. Steve drove 50 km himself.

D. You want to find the total distance.

Possible guidelines for assessment items—thinking skill 7. One way to assess this skill is to have students choose the reason why a given answer is reasonable. Another is to have them choose the reason why a given answer is unreasonable. Items 7.1, 7.2, and 7.3 illustrate these methods. However, asking children why an answer is *not* reasonable may be confusing (item 7.1), and you may wish to omit that tactic.

Completion Tests

What are completion tests?

A *completion test* is made up of items that are answered by supplying requested information. The information may be simply a word, number, or phrase that will correctly complete a statement, or it may be a sentence or collection of symbols that satisfy the request.

Completion items are useful for evaluating the ability to get a correct answer and the ability to use problem-solving thinking skills. In this section, we will focus on using them to measure the seven problem-solving thinking skills described earlier. For example, the following is a completion item that is designed to measure a student's ability to recognize needed data.

Problem-solving thinking skill. Select or find data needed to solve the problem.

Objective. Given a problem with unneeded data, identify the data needed to find a solution.

Completion item. What data in this problem would you use to find the solution?

Ted works 9 hours each day. He receives 21 days of vacation each year. How much does he earn in a 25-day work month if his pay is $7.65 per hour?

Give it a try!

Change the problem below to include extra data and use it to write a completion item like the one above that measures the student's ability to decide which data are needed to solve the problem:

Problem: You have $30 to buy birthday presents for friends. If you buy a book for $9.85, a record for $7.69, and a pen for $6.58, how much will you receive back in change?

What are some advantages of using a completion test?

Advantages of this method of evaluation include the following:

- It is one of the easiest tests to construct.
- It requires the student to supply the answer, thus providing less reward for guessing.
- It allows for viewing students' work, thus providing a greater understanding of their problem-solving processes.

56

What are some disadvantages of using a completion test?

Disadvantages include the following:

- It is a difficult test to grade, since different interpretations and partial answers are possible.
- It requires more time to administer, since more writing is involved.

When and how should completion tests be used?

Completion tests can measure a student's ability to find the correct answer to a problem and to use certain problem-solving thinking skills. They are more useful than multiple-choice tests for analyzing a student's procedure for solving a given problem, especially if you require students to show their work as well as give the answer. Completion tests are most useful in classroom testing situations where the tests are graded by the teacher. When your students take a completion test, allow them plenty of time to think through the situation or problem and write their responses.

Completion items should not be the exclusive means for evaluating problem solving. They can be useful in analyzing a student's procedure for solving a given problem and gaining specific insight into a student's ability to use higher level problem-solving thinking skills, but other techniques can provide more depth in these situations.

How do you develop completion items to measure problem-solving thinking skills?

When you prepare a completion test, first specify exactly what ability you want to measure and prepare the items so that they do, indeed, evaluate this ability. Use the following procedure:

1. Identify the outcomes you want to measure.
2. Prepare specific objectives for these outcomes.
3. Write items to measure outcomes in terms of the specified objectives.

The validity of an item can be assessed intuitively by a careful analysis of exactly what the item requires the student to do or know. Valuable information is gained by asking students and other teachers to participate in this analysis.

Sample Completion Items

To illustrate how completion items can be used to evaluate a student's ability to use the seven problem-solving thinking skills given earlier, we have included ten sample assessment items together with guidelines for writing additional items. One item is given for each objective for each problem-solving thinking skill. Strategies used for solving the problems, content, types of numbers, and sources of data are also varied in these sample problems.

Problem-solving thinking skill 1. Understand or formulate the question in the problem.

Objective. Given a problem, select, write, or state in your own words the question that would be answered when a solution is found.

Completion item 1. Rephrase the question in this problem in your own words:

Eric weighed 147 pounds. How much did he weigh after he had lifted weights, eaten more, and gained 15 pounds?

Possible answer: How much does Eric weigh now?

Problem-solving thinking skill 2. Understand the conditions and variables in the problem.

Objective. Select and identify the key conditions and variables useful in understanding and solving the problem.

Completion item 2. List two important conditions that should be kept in mind when solving this problem:

A T-shirt shop had only three types of iron-on digits, 1, 3, and 8, left to make numerals on shirts. If digits can be repeated, how many different 2-digit numerals can they make?

Possible answer: 1. The only digits that can be used are 1, 3, and 8. 2. Digits can be repeated.

Problem-solving thinking skill 3. Select or find data needed to solve the problem.

Objective. Given a problem with unneeded data, identify the data needed to find the solution.

Completion item 3a. Which of the data given (possibly all) are needed to solve the problem?

Jack weighed 43.6 kilograms. He weighed 19.8 kg more than Jennifer. Tim weighed 12.7 kg more than Jennifer. How much did Jennifer weigh?

Possible answer: Jack weighed 43.6 kg; Jack weighed 19.8 kg more than Jennifer.

Objective. Given a problem with missing data, describe or find appropriate data needed for solving the problem.

Completion item 3b. What additional data, if any, are needed to solve this problem?

58

An airplane's flying speed without wind was 843 km/hr. Its speed was increased by a tailwind. What was the resulting speed?

Possible answer: The tailwind speed was 65 km/hr.

Completion item 3c. Which data must be found in the table to solve this problem?

Joe ate a cup of fruit that provided 200 calories. What fruit did he eat?

100 Calorie Portions of Selected Foods

Fresh orange juice	9/10 cup
Canned apricots	1/2 cup
Canned peaches	3/5 cup
Canned corn	3/4 cup

Possible answer: We see in the table that 1/2 cup of canned apricots provides 100 calories, so a cup would be 200 calories.

Problem-solving thinking skill 4. Form subproblems and select an appropriate solution strategy to pursue.

Objective. Given a multiple-step or process problem, formulate or select subproblems that could be solved to find the solution.

Completion item 4. Write two problems that can be solved to help find the solution to this problem:

Sue bought 6 pairs of socks. Each pair cost $2.75. She gave the clerk a $20 bill. How much change did she get back?

Possible answer: 1. What is the cost of 6 pairs of socks at $2.75 a pair? 2. If you buy socks for $16.50, how much change do you get back from a $20 bill?

Problem-solving thinking skill 5. Correctly implement a solution strategy and attain subgoals.

Objective. Given a story problem, write a number sentence that could be used to solve that problem.

Completion item 5a. Write a number sentence that could be used to solve this problem:

A stock gained 7/8 of a point on Wednesday. It gained 4 times this much on Thursday. How much did it gain on Thursday?

Possible answer: $4 \times 7/8 = 3 1/2$

Objective. Given a story problem, select or draw a picture that could be used to help solve the problem.

Completion item 5b. Draw a picture that could be used to help solve this story problem.

Clinton, Springfield, and Weldon are the corners of a triangle. It is 45 miles from Clinton to Springfield, 49 miles from Springfield to Weldon, and 12 miles from Clinton to Weldon. How much farther is a trip from Springfield to Weldon through Clinton than a trip directly from Springfield to Weldon?

Possible answer:

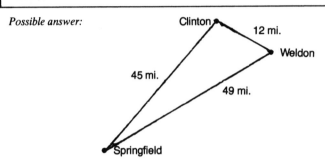

Problem-solving thinking skill 6. Give an answer in terms of the data given in a problem.

Objective. Given the numerical part of the answer to a problem, write the answer in a complete sentence.

Completion item 6. Here is a problem and the numerical part of its solution. Write the answer using a complete sentence.

A farmer had chickens and pigs in his barn lot. How many chickens and how many pigs did the farmer have if he counted 20 heads and 56 feet? Numerical part of the solution: 12, 8

Possible answer: The farmer had 12 chickens and 8 pigs.

Problem-solving thinking skill 7. Evaluate the reasonableness of an answer.

Objective. Given a problem and its answer, estimate to decide if the answer is reasonable.

Completion item 7. A problem and its answer are given. Estimate and decide if the answer is reasonable.

A set of wrenches priced at $3.95 each costs $98.75. How many wrenches were in the set?
Numerical answer: 19

Possible answer: 20 × $4 is only a total of $80. The answer 19 is not reasonable.

These sample items suggest ways to evaluate the problem-solving thinking skills that are important for students. When writing similar items, keep the following guidelines in mind:

1. Be sure the item really measures whether or not the student can do the skill.

2. Write the item so that the question is clear and the requested answer is as definite as possible.

3. Use problems, pictures, tables, and so forth in the items to make the answer as unambiguous as possible.

4. Vary the characteristics of the problems used in the items (problem type, type of numbers, source of data, and so forth).

3

How Do You Organize and Manage an Evaluation Program?

NOW that you have seen a variety of evaluation techniques for problem solving, the next step is to consider how to incorporate these techniques into your mathematics program. In this chapter we discuss three topics:

1. Beliefs about designing an evaluation program
2. Guidelines to assist in making decisions about an evaluation program
3. Sample evaluation programs for "real" classrooms

Before beginning our discussion of these topics we think it is important to reiterate a point that underlies everything we have considered so far. That is, *for students' problem-solving abilities to improve, they must believe that you regard problem solving as important.* If students sense that problem solving is important to you, they are likely to view it as important as well. A key ingredient in imparting this sense of importance is to evaluate students' problem-solving progress on a regular basis.

BELIEFS ABOUT EVALUATION PROGRAMS

Before developing an evaluation program, you need to specify what beliefs you have about problem solving. That is, you should have a clear idea in your mind about the "givens" that will direct your program. We did this. We identified several beliefs that form the basis for everything that appears in this book. We think they can serve you also as you begin to devise your own program.

Our first belief is that any evaluation program must be based on a well-conceived plan. Develop your evaluation plan at the same time you are developing your instructional objectives and activities. Linda, the teacher in our opening dialogue, had a carefully developed set of classroom problem-solving activities but no evaluation plan. Don't make the same mistake!

Secondly, we believe that real growth in problem-solving ability often takes place rather slowly over a prolonged period of time. Don't be too ambitious or set your goals too high. In the beginning, you may be content with looking for changes in students' attitudes and beliefs, two essential ingredients for later problem-solving success.

Our third belief is one that we share with every experienced teacher. That is, students' performance can be influenced greatly by the constraints imposed by the evaluation. It is important to make evaluation as nonthreatening for students as possible.

Our fourth belief is related to the third: Individual, informal interviews are the most valid method to assess students' thinking processes. Brief, casual, one-on-one interviews should be held whenever feasible.

The attention we've given throughout this booklet to evaluating thinking processes should not be interpreted as an indication that getting correct answers is undesirable. On the contrary, our fifth belief is that the ability to get correct answers more often should be an important goal of problem-solving instruction.

Finally, we believe that among the purposes for evaluating is to have a basis for instructional decisions with respect to the effectiveness of instruction, the identification of students' strengths and weaknesses, and assigning grades. Thus, assigning grades is only a part of the overall aim of evaluation; much more is involved.

With this set of beliefs in mind, we present several guidelines, or "evaluation heuristics," that we recommend for your consideration as you begin to develop your own evaluation program.

GUIDELINES FOR IMPLEMENTING AN EVALUATION PROGRAM

The guidelines we have devised have evolved from our collective experiences over the years in designing mathematics curriculum materials, from our analysis of existing evaluation materials, and from our judgments concerning the ingredients of a successful problem-solving program.

The development of these guidelines was no simple matter. Two factors made the task difficult. First, we established a wide variety of goals for teaching problem solving, ranging from the acquisition of particular kinds of skills, such as the ability to make a table, to the development of attitudes conducive to good problem-solving performance, such as a willingness to persevere in problem-solving efforts. Thus, our desire to keep the number of guidelines small was hampered by the complexity and diversity of the instructional goals. A second factor that made our task difficult was that many of the goals are not amenable to assessment, at least not through traditional means. For instance, how does one assess perseverance? Nevertheless, we have decided on a reasonably small number of guidelines that should be helpful as you begin to plan your own program for evaluation.

1. Evaluate your students' work on a regular and systematic basis.

2. Evaluate their thinking processes as well as their answers.

3. Match your evaluation plan to your instructional goals.

4. Assess attitudes and beliefs about problem solving as well as performance.

5. Observe students' small-group efforts and their written work as an important part of your evaluation plan.
6. Interview students individually whenever possible.
7. Do not feel compelled to evaluate all students at the same time or to record their performance on every problem-solving experience.
8. Advise students of your evaluation plan and how it works.

A few comments about these guidelines are warranted before we present some sample evaluation programs. First, it is essential that you establish an evaluation routine in your classroom. That is, your students should learn to accept the idea and expect that you will formally assess their work regularly (perhaps once a week) and that you will also conduct informal assessments of their work continually through observations of their group work, brief chats, and so forth. Strive to make evaluation a natural, positive event. These informal procedures will probably be your best source of information about your students' attitudes, frustrations, anxieties, and beliefs. Often they will also be your only means for getting a good idea about how they are thinking as they solve problems.

Guideline 7 deserves special comment. Evaluation need not be an end-of-the-unit event. Sure, you can give a whole-class test at the end of a specified period of time, but don't limit your evaluation to this approach. Remember, you are looking for changes (positive ones) in students' performance and you want to observe these changes as they occur, not hope they occur when you want them to. Consequently, on any given day you will probably be able to gather evaluation data from only a few of your students (perhaps as few as one or two). However, over time you can observe all your students several times, and these observations, taken together with interviews and samples of written work, will be ample information for evaluation purposes.

Finally, once you have decided on an evaluation plan, share it with your students. Let them know that you think it is important for them to do their best to reach the goals you have set for them. An open discussion about what you expect of them and how you intend to measure their performance is a key to a successful evaluation program.

SAMPLE EVALUATION PROGRAMS

To give you a clearer picture of how to implement an evaluation plan, we describe in this section the evaluation programs of three teachers. Although these programs are hypothetical, our experience suggests that they are realistic approximations of the sort of program you could put in place in your own classroom. Before you begin to consider the merits of these samples, you should note that they were created with certain important classroom realities in mind.

Realities of Classroom Evaluation

Perhaps the most limiting reality of day-to-day classroom life is *time*. There simply is not enough time to do all the activities you would like to do or to cover all the topics in the depth you would like to cover them. Furthermore, it is likely that the noninstructional demands on your time have increased since you first began teaching. Consequently, we recognize that you probably will not have the time to permit individual student interviews to be your primary assessment technique. However, we do believe that you can incorporate brief, spontaneous "chats" with individual students that will give you a wealth of information about their growth as problem solvers.

A second classroom reality that may influence the type of plan you design is that you are probably required to give your students grades. If so, you will need to decide what proportion of the mathematics grade should be based on problem-solving performance. We suggest that you strike a balance among three areas: conceptual understanding, the acquisition of skills, and problem-solving performance. For a particular grading period, you might choose to give more weight to one of these areas than the others, or you may decide to give them equal weight. Whatever you decide, keep in mind the underlying theme for evaluation: If you want your students' problem-solving abilities to improve, you must evaluate their progress on a regular basis and use the results to direct subsequent instruction.

Finally, and possibly most importantly, in the final analysis you are your own best evaluation instrument. If our personal judgment does not jibe with the results from one evaluation source, by all means trust your judgment or use other evaluation techniques. Your teaching intuition should not be ignored in your effort to assess your students' growth.

Program 1: A Grade 2 Classroom

This is Ruth's fifth year as a second-grade teacher. She is a firm believer in the importance of establishing a foundation in the early grades for concepts taught in the upper grades, but she has never before attempted to make problem solving a key part of her math program. She also recognizes the value of establishing a classroom climate that is conducive to problem solving—that is, she wants her students to enjoy learning and in particular to enjoy solving problems. Using these beliefs as guides, she has set four goals for her students:

1. Demonstrate a willingness to engage in problem solving.
2. Develop self-confidence in their abilities to solve problems.
3. Be able to identify the important information in a problem statement.
4. Be able to use the following strategies: choose the operation, draw a picture, look for a pattern, use objects, and guess and check.

Her attitude is that if her students attain these goals, they will be prepared for future problem-solving experiences. Notice that two of her goals (1 and 2) are primarily *affective* in nature, and the others are related to the actual

66

process of solving problems. Because she likes to have her children work cooperatively in groups of four on many of their problem-solving activities, Ruth is able to gain a considerable amount of information about their problem-solving attitudes and behaviors by observing and listening as she walks around the room. When the twenty children in her class are working in groups of four she has found it fairly easy to observe the behavior of all of them.

Ruth has decided to use three evaluation techniques: (1) general impression scoring, (2) a problem-solving observation rating scale, and (3) readiness-skills inventories. She has a problem-solving folder for each child in which she keeps rating scales and skills inventories. About once a week she collects a sample of each child's work on a problem and assesses the quality of the work based on a rather quick look at the paper. She does not use this "general impression" to assign grades but rather as a means for gaining a general sense of the childrens' overall progress.

Ruth developed the observation rating scale in figure 23 to reflect the four goals she had already established. Her observations of the children's group work, together with her general impressions from looking at individual papers, serve as her sources of data for completing the scale. Of course, she does not complete a scale sheet for every student every day. To attempt such a thing would be both highly impractical and unnecessary. Instead, she tries to complete one scale for every student about every two weeks. By so doing, in the course of about two months' time she has three or four sheets for each child. This is ample information on which to judge a student's progress. On a particular day she will decide to focus her attention on one group (without ignoring the others, of course). At the end of the day she is often able to complete most of the scale for all four children in the group.

Student: _____ **Date:** _____

Frequently Sometimes Never

1. Shows a willingness to try problems
2. Demonstrates self-confidence
3. Selects all important information
4. Uses strategies appropriately:
   a) choose the operation
   b) draw a picture
   c) look for a pattern
   d) guess and check
   e) use objects

Fig. 23. Ruth's Problem-solving Performance Rating Scale

The readiness-skills inventories are used to help Ruth pinpoint specific areas of strength and weakness. Twice a week she devotes five to ten minutes to readiness activities. For example, she may show the class a picture and ask them to make up stories about the picture that involve addition; or she may read a story problem and ask them to tell her which operation could be used to solve the problem. (For a more complete discussion of readiness activities, see Charles and Lester [1982].) About once every two weeks she collects the children's work on a readiness activity and grades it in terms of the extent to which a particular skill was used correctly. Once again, the scoring of these twenty papers does not take an inordinate amount of time because only a single example of a single skill is involved.

Program 2: A Grade 5 Classroom

Nick began to develop a systematic problem-solving program for his fifth graders two years ago. From the beginning, he has been interested in helping his students attain a wide range of goals, both cognitive and affective. His problem-solving program has the following characteristics: (1) it includes several different problem types, such as one-step and multiple-step story problems, process problems, and real-world applied problems; (2) it places emphasis on developing the various thinking processes associated with good problem solving; (3) it gives considerable attention to the use of several key strategies, such as *guess and check, look for a pattern, solve a simpler problem,* and *work backward*; (4) it stresses the value of perseverance, willingness to take risks, and sharing ideas with others; and (5) it stresses the development of certain fundamental beliefs about solving problems, for instance, most problems can be solved in more than one way and some problems have more than one correct answer.

Only now, in his third year of making problem solving a major focus of his instruction, has Nick begun to feel comfortable with the evaluation techniques he has designed and used. This year he chose to begin by relying on three evaluation methods: (1) an analytic scoring scale, (2) an observation checklist, and (3) a strategy inventory. Like Ruth, our second-grade teacher, he has a problem-solving folder on each student. Also, like Ruth, he collects samples of his students' work on a regular basis.

Before collecting a set of papers, Nick decides on certain students (no more than four or five, and often students who have been working together) whose work will be given close attention. All students' papers will receive a check mark indicating that work was completed but, in addition, the papers of students identified for special scrutiny will be evaluated using an analytic scale much like the one described in this book. At the top of these papers he writes an ordered triple of numbers corresponding to the three categories of the scale. He also writes "NG" to indicate that *no grade* is to be associated with the scores (a paper might have 2, 1, 1 NG). He likes this approach because it gives him data in three important areas (understanding, planning, and answering) while at the same time keeping his students aware that he looks carefully at their entire problem-solving effort, not just their answers. Furthermore, by telling them that no grade is being assigned to their work, he can give them feedback on their performance in a relatively nonthreatening way.

The analytic scale also helps Nick assign grades. Once every grading period he uses it to evaluate his students' work solving one-step, multiple-step, and process problems. Scores on these problems are recorded in the students' problem-solving folders and are used in conjunction with other evaluation results to determine math grades for the grading period.

Additional information on students' performance and attitudes is obtained by means of an observation checklist (fig. 5). Nick's method is to complete as much of the checklist as possible for each of the students who were singled out. A completed checklist reflects a composite of Nick's analysis of the student's written work and his informal observation of the student's small-group behavior.

Finally, about once every three or four weeks Nick asks his students to complete a self-report on their use of strategies (fig. 11). This inventory is completed immediately after the students have completed work on a process problem. It provides supplemental data about his students' awareness of, and beliefs about, various problem-solving strategies. It also allows students the opportunity to make personal commentaries about a particularly important aspect of problem solving. From past experience Nick has learned that this inventory will not be needed after the first four months of school, since by that time students have typically become quite aware of the value of having a good repertoire of strategies.

Summary data from the analytic scales, observation checklists, and self-report inventories are kept for each student in a problem-solving evaluation notebook. Nick has learned that the data in this notebook will be invaluable to him in making decisions about instruction throughout the year.

Program 3: A Grade 7 Classroom

Marla has always emphasized problem solving in her seventh-grade classroom. In her eleven years as a middle school teacher she has firmly believed that the main reason for learning mathematics is to be able to solve problems better. Consequently, she has been an avid proponent of incorporating problem-solving as a major part of her evaluation program. For each of the past four years, Marla has taught at least one math class of low-achieving students. She has found that these students not only are lacking in the mastery of computational skills but, more importantly, are weak in several fundamental problem-solving skills, notably the ability to select data needed to solve verbal problems, the ability to organize information, and the ability to carry out a plan of attack on a problem to completion. Furthermore, due to their histories of poor performance, most of them have very negative attitudes toward mathematics and little or no confidence in their ability to do math. For these reasons she has decided to pay special attention to developing certain problem-solving skills, changing attitudes, and increasing self-confidence.

69

The evaluation plan that Marla has decided on includes three techniques not considered by either Ruth or Nick. She will employ a focused holistic scoring scale, problem-solving skills inventories, and small-group interviews. In addition, she will be able to get valuable information as to her students' skill development, attitudes, and self-confidence by observing them as they work in groups of four (the grouping scheme she has used faithfully for several years).

Once a week she plans to collect students' written work on one or two problems. She will use the focused holistic scale (fig. 16) to assign scores to each student's paper. The class will be told how the scale works and that the scores will not be used for grading purposes but rather as a means of helping them keep track of their own progress. About twice a week she will have each student complete a skills inventory lasting no more than five minutes. A typical inventory will contain two items such as the following:

This problem has missing data. Make up needed data and solve the problem.

Bart and Hanna put 2356 beans in a jar for a math contest. George's guess was closest. By how much did George miss the total?

Tell which operations ($+ - \times \div$) you can use to solve this problem.

The U.S. Congress has □ senators and □ representatives. There are □ more members of Congress than there are judges on the U.S. Supreme Court. How many judges are on the Supreme Court?

Decide if the answer to this problem makes sense. If it does not make sense, tell why.

Problem: A rectangular field is 97 meters wide and 133 meters long. What is the area of the field?

Answer: 230 square meters

Marla has developed a bank of over seventy skill items corresponding to the seven categories of thinking skills discussed in chapter 1, "What Are You Trying to Evaluate?" She plans to keep a record of each student's performance in each category. She can use this information to make decisions about areas of needed instruction and to help her assign grades.

Perhaps Marla's most useful assessment method will be the regular interviews she will have with her students. Usually she has twenty-four to twenty-eight students in her class, so she will have six or seven groups of four. Each week she will have an interview (or "chat," as she calls it) with two or three groups, spaced over two or three days. While she is talking with a group, the rest of the class will be engaged in some sort of independent activity (either in groups or as individuals). A typical chat will last about fifteen minutes and will be devoted to a discussion of how well the group has been working together, areas in which the group feels improvement is needed, and so on. During this time Marla will also try to learn more about the students' attitudes and beliefs about math in general and problem solving in particular.

The evaluation plans of Ruth, Nick, and Marla illustrate the wide variety of possible approaches that can be employed. We expect you will want to choose from among the several techniques described here and design a plan that best fits your classroom situation.

4

How Do You Use Evaluation Results?

THE primary reason for developing and using an evaluation plan for problem solving should be to gain information that enables you to make instructional decisions based on the identified strengths and weaknesses of your students. In this section we share some ideas about the kinds of instructional decisions you may need to make when you evaluate your students' progress in problem solving. We discuss these decisions in three areas: the classroom climate for problem solving, the content of instruction and teaching methods, and assigning grades. Let us look at the first.

CLASSROOM CLIMATE

A classroom climate conducive to problem solving is essential to building a successful program. Evaluation techniques such as attitude inventories, student self-reports, and student observations can provide data necessary to make judgments about students' attitudes and beliefs. If assessment data suggest that their attitudes and beliefs are not positive, there are several sources you should examine as possible causes. One of these could be the content of your program. Are problems appropriate with respect to level of difficulty? The problems you use should not be always easy or always difficult. Strive, instead, for a variety of experiences that span a range of difficulty. Are experiences sequenced to develop students' skills and confidence gradually? In general, your curriculum should move from easier to more difficult experiences. Necessary problem-solving skills and strategies should be taught, followed by experiences that sharpen students' decision-making skills

A second source to examine for negative attitudes among students is the time commitment you have made to problem solving. Is problem solving a regular and frequent part of your program, or do students consider it an "extra"? If problem solving is not a regular and frequent part of the instructional program, students will not consider problem solving to be that important and their problem-solving abilities will not improve as much as you would like.

A third point for examination is your evaluation practices. Is student performance graded in every problem-solving experience? Constant grading can negatively affect the problem-solving atmosphere. Do you evaluate more than the answer? Focusing only on the answer will usually have an adverse effect on students' attitudes and beliefs about problem solving.

Fourth, examine your own attitudes, beliefs, and teaching methods. Are you excited about problem solving? If you are enthusiastic and communicate the importance of problem solving to your students, they will enjoy it and exert the effort needed to improve their skills. Do you provide the appropriate amount and kind of assistance your students need? Students who are stumped often need guidance from the teacher to avoid excessive frustration, particularly those for whom problem solving is new.

Here are some additional ideas that can promote a positive environment.

1. Encourage students to contribute problems from their personal experiences.
2. Personalize problems whenever possible; for example, use students' names in problems.
3. Recognize and reinforce willingness.
4. Reward risk takers.
5. Encourage students to play hunches.
6. Accept unusual solutions.
7. Praise students for getting correct answers, but emphasize their selection and use of problem-solving strategies.
8. Emphasize persistence rather than speed.

CONTENT OF INSTRUCTION AND TEACHING METHODS

The second area in which evaluation data can be used to make instructional decisions is the content of your problem-solving program and your teaching methods. Evaluation data from such sources as observations, interviews, and analyses of written work can be used to diagnose students' strengths and weaknesses relative to the thinking processes involved in solving problems. The precision with which you can diagnose strengths and weaknesses is influenced by the evaluation techniques you use. For example, individual student interviews will usually give you a more detailed diagnostic profile of a student's thought processes than analyzing their performance on a multiple-choice test.

We have found the following four areas general enough to pinpoint strengths and weaknesses reliably yet specific enough that subsequent instruction can be prescribed. The areas are (1) understanding the problem; (2) developing a plan; (3) implementing the plan; and (4) answering the problem and checking the results. Following is a list of ideas to consider with regard to changes in the content and teaching methods when student weaknesses are detected in these four areas.

Understanding the problem

1. Have discussions that focus on understanding the problem before students start work on it.
   a) Ask questions that focus on (1) what it is they are asked to find (i.e.,

the question), (2) the conditions and variables in the problem, and (3) the data (needed and unneeded).

b) Have students explain problems in their own words.

c) Remind them of similar problems.

d) Have them use colored markers to highlight important phrases and data.

e) Have them list the data given.

2. Develop students' skills at understanding the question and data in a problem by using activities such as these:

a) Give them a story, and have them write a question that can be answered using data in the story.

b) Give them a problem that includes unneeded data and have them identify the data needed to find a solution.

c) Give students a problem with missing data and have them find appropriate data so the problem can be solved.

Developing a plan

1. Suggest a solution strategy with the problem statement.

2. Discuss possible solution strategies before students start solving a problem.

a) Have them suggest the reasons they believe particular strategies might work.

b) For one-step and multiple-step problems, have them tell what action is taking place that suggests a particular operation.

3. Remind students of similar problems.

4. Use activities such as these that can improve students' abilities to select a plan:

a) Give them a one-step or multiple-step problem. Have them tell the operation or operations needed to find a solution.

b) Give them a one-step or multiple-step problem without numbers. Have them tell the operation or operations needed to find a solution.

c) Give students a completed solution to a problem (such as a number sentence, an organized list, or a picture). Have them think of a problem that would fit the solution.

5. Discuss solution strategies used in solving a problem after students have completed work on the problem.

a) Have them tell why they selected particular strategies.

b) Show different solutions (or strategies), if possible.

c) Evaluate the usefulness of different solution strategies after students have completed work on the problem.

d) Show and discuss incorrect solution attempts (or inappropriate strategies) that have been used by students. Discuss which strategies were used and why they were not appropriate.

Implementing the plan

1. Have students evaluate the implementation of a solution strategy to determine whether it was done accurately.
2. Give the start of a solution (or a strategy) and have them complete the solution to find the answer.
3. Give a hint with the problem statement (such as a strategy) telling how to start a solution.
4. Give direct instruction and practice with particular solution strategies.
5. If possible, show solution strategies which, if properly implemented, would have led to the correct solution to the problem. Show where the error occurred in implementing the strategy.

Answering the problem and checking the answer.

1. Have students check to be sure they used all important information in the problem.
2. Have them check their arithmetic.
3. Have them answer problems in complete sentences.
4. Use activities that develop students' abilities to give the answer to a problem and check their work.
   a) Have them estimate to *find answers.*
   b) Have them estimate to *check answers.*
   c) Give them the numerical part of an answer, and have them state the answer in a complete sentence.
   d) Give them a problem and an answer. Have them decide whether the answer is reasonable.

ASSIGNING GRADES

The third area involving instructional decisions is concerned with assigning grades. A grade can be assigned to pupils' progress in problem solving when it is necessary to document their progress. However, remember that *evaluation* is not synonymous with grading. Every teacher should have a plan for evaluating progress in problem solving whether or not grades are assigned.

If you choose to, or must, assign a grade, here are some guidelines to consider.

1. Use a grading system that considers the process used to solve problems, not just the answer. If your instructional program emphasizes the process of solving problems (as we believe it should), then your grading system should be consistent with this emphasis. The sample holistic scoring systems described earlier will help you evaluate problem-solving performance in terms of both the process and the answer.

2. Advise students in advance when their work will be graded. They should not participate in problem-solving experiences wondering whether or not their work will be graded.

3. Be aware that pupils may not perform as well when they are to be graded. A "testing" situation usually has a negative influence on performance.

4. Use all available evaluation data as a basis for assigning grades. Avoid assigning a grade based on one measure. Use all the formal and informal evaluation data at your disposal to make decisions regarding grades.

5. Consider using a testing format that matches your instructional format. For example, if cooperative learning groups play a major role in your program, consider testing performance in cooperative learning groups. Each student in the group receives the same grade for the "official" group paper that is turned in.

As the role of problem solving in the mathematics curriculum continues to grow, the need to assign grades may also grow. For many students, grades can be a positive motivation, particularly when the system used to assign a grade reflects the many facets of problem-solving performance.

Some Guidelines for Using This Book for In-Service Education

WE TRIED to write this book so it can be used as a "self-instructional" program. That is, if you were to read this book and do the "give it a try" activities, you would end with a level of understanding sufficient to develop and begin using your own evaluation plan for problem solving. Many of you who read this book may at some time be responsible for educating others on the topic of problem-solving assessment. Therefore, we think it might be helpful to share some of our experience in educating teachers on this topic.

We'll assume that teachers participating in the in-service program have had previous in-service education related to problem solving. In particular, we'll assume they are familiar with the problem-solving curriculum and techniques for teaching problem solving.

Most teachers know that problem-solving assessment is a complex activity, but they have not thought seriously about how to address it. They know that a check for the correct answer is too restricted, but what to do is not clear. Therefore, in planning an in-service program, we organize our activities into four program phases:

1. Make teachers aware of possible goals for problem-solving instruction and sensitize them to the need for problem-solving assessment.
2. Sensitize teachers to the complexities of problem-solving assessment.
3. Introduce assessment techniques.
4. Help teachers develop their own assessment plans.

We'll now share some of our thoughts on how you can use this book to implement these phases of an in-service program. (We've made no assumptions about the time available for in-service training. Like most in-service programs, opportunities to try the ideas with students and then return to discuss those experiences with other teachers best promote the development of teachers' assessment skills.)

Make teachers aware of possible goals for problem-solving instruction and sensitize them to the need for problem-solving assessment.

We often find that many teachers attending an in-service program on evaluation are somewhat familiar with goals for teaching problem solving. Nevertheless, we begin most in-service programs on problem-solving assessment with some discussion of goals for problem-solving instruction. We begin with a small-group activity where teachers generate goals for problem-solving instruction. Then each group shares its goals, and we list them on the chalkboard. We then introduce the goals given in the first part of this book and compare them to the list on the board. We use this as an opportunity to emphasize two important points. One is that problem-solving instruction has many important goals beyond developing students' abilities to get more correct answers. Goals related to students' thinking processes, attitudes, and beliefs are also important. The second point we make is that an assessment program for problem solving must be matched to the instructional goals of the program. Most realize at this point that an assessment program for problem solving must go beyond a check for the correct answer.

Sensitize teachers to the complexities of problem-solving assessment.

As we said above, most teachers are aware that problem-solving assessment is a complex activity, but they have not tried to formulate in their minds the issues involved. We begin this phase of the in-service program with an activity that gets teachers thinking about assessment issues. One way we do this is to have teachers discuss students' written work on a problem. The two solutions given in figures 18 and 19 to the Ping-Pong tournament problem or the ones in figures 13–15 for the chickens and pigs problem can be used for this activity. After clarifying a correct solution to the problem, we allow time for teachers to discuss each student's solution, trying to come up with ideas about the thinking the student used to generate the solution. Discussion about the students' thinking and how the teachers would evaluate performance on these papers is usually quite lively. We end the discussion by sharing the following points about evaluation and problem solving:

- Evaluation is not synonymous with grading.
- We should evaluate thinking processes as well as the correct answer.
- We should always attempt to observe and question students while they solve problems.
- We should match our assessment plan to our instructional emphases.
- We should assess attitudes and beliefs as well as performance.
- We should try to interview students on occasion.
- Every student does not have to be evaluated in every problem-solving experience.
- Students should be informed of the teacher's evaluation plan.

This activity helps most teachers realize the complexity of problem-solving assessment and helps clarify many of the issues involved. Teachers now often ask something like, "If problem-solving assessment is this complex, what can I do in my class of thirty students!?" They are now ready to explore assessment techniques.

Introduce assessment techniques.

This phase of the in-service program can last from a few minutes to several hours depending on the number of techniques introduced and the extent of practice with each technique. We make an effort to address two areas, as a minimum, in every in-service program—techniques for observing and questioning students, and holistic assessment techniques. We selected these areas because we strongly believe that every teacher's evaluation plan should include the observation and questioning of students as they solve problems (at least informally), and most teachers faced with thirty students need to collect and examine students' written work, since they do not have the time to interview students individually as they solve problems.

The ideas included in this book for observing and questioning students are difficult to practice in an in-service setting. Given sufficient time, peer-group teaching can be used as a setting to practice these techniques. We often have teachers practice writing various kinds of questions that would help them assess students' thinking for particular problems. Practice in writing assessment questions followed by an opportunity to try those questions with a group of students followed in turn by additional discussion with peers is a process we have found successful in developing teachers' skill in asking questions for assessment purposes.

The holistic scoring schemes given in this book are easily practiced in an in-service setting. We usually begin with the analytic scale and have teachers work in groups evaluating the pupils' papers discussed in the second phase of the program. As time permits, we provide additional practice in evaluating student solutions to other problems. We next practice using the focused holistic system, using at least the same student papers used for practicing the analytic scale. We follow this with a discussion of the advantages and disadvantages of the two schemes and suggestions for situations in which each might be used. We always encourage teachers to suggest ways to modify or extend the holistic evaluation scales to make the scales most useful for particular situations (e.g., for primary grades versus intermediate grades). Often these ideas lead into the techniques discussed in the section in this book on assessment data from students (e.g., having students in the upper grades write a short paragraph explaining how they arrived at their solution). We end the discussion on holistic scoring by discussing general impression scoring, emphasizing that this technique should be used only by those experienced in the teaching and assessment of problem solving.

Next we explore techniques for gathering assessment data from students and from multiple-choice and completion tests. We introduce ideas in these categories in the same order in which they are given in this book. We usually do not spend much time with teachers on writing multiple-choice test items. Instead, we try to spend time giving teachers practice writing completion assessment items like the samples given in this book. Practice in writing completion assessment items is a lively activity, challenging to many teachers, and particularly useful for both instruction and assessment. (Many completion assessment items can also be used for instruction on problem-solving skills.)

After teachers are exposed to a variety of assessment techniques, it is not uncommon for them to feel overwhelmed. They now know there are several techniques they can use, but they also know they cannot possibly use all of them. This awareness motivates them to deal with the practical issue of building their own evaluation plans.

Help teachers develop their own assessment plans.

We begin this phase of the program by reviewing the sample evaluation plans given in this book and discussing their advantages and disadvantages. We next talk about ways to modify these hypothetical plans. Now teachers are ready to build their own plans, typically working in small grade-level groups to do this. (Ideally, the number of teachers is large enough for several same-grade-level groups to be formed.) We next share and discuss the teachers' plans. This activity clarifies further the nature of some of the evaluation techniques they have learned and provides ideas for additional ways to modify these techniques.

The four phases discussed above prepare teachers to begin the real development of their evaluation plans. Once a teacher has developed a plan in an in-service program, the next step is to try it and subsequently revise it to a workable plan.

We have said several times in this book that we believe the primary reason for using an assessment plan is to collect data that enable the teacher to make instructional decisions. A final phase can be added to the in-service program to deal with the analysis and interpretation of evaluation data and the planning of subsequent instruction based on these data. We have found that this activity is most useful after teachers have some experience teaching problem solving and implementing their evaluation plans. Teachers returning to an in-service program are then ready to address instruction as a follow-up to assessment. The chapter entitled "How to Use Evaluation Results" can be used in this kind of activity.

Further Reading on Mathematical Problem Solving

Burns, Marilyn. *The I Hate Mathematics Book*. Boston: Little, Brown & Co., 1975.

This book contains a variety of interesting and challenging mathematics problems aimed at convincing kids that they really do like mathematics. Written for students and teachers.

———. *Math for Smarty Pants*. Boston: Little, Brown & Co., 1982.

Similar to the one above, this book discusses a variety of interesting and challenging problems. Written for students and teachers.

Charles, Randall, and Frank Lester. *Teaching Problem Solving: What, Why and How*. Palo Alto, Calif.: Dale Seymour Publishing Co., 1982.

This book describes a district-tested problem-solving program for grades 1–8. Detailed directions are given on how to organize, implement, and evaluate a problem-solving program for the classroom.

Charles, Randall, and others. *Problem Solving Experiences in Mathematics, Grades 1–8* (eight books). Menlo Park, Calif.: Addison-Wesley Publishing Co., 1985.

Each of the eight books provides teachers with a problem-solving experience for almost every day of the school year. Problem-solving skill activities, one-step problems, and process problems are included in each book. Multiple-step problems begin in grade 3. A teaching strategy in a lesson-plan format is provided.

Cook, Marcy. *Think about it! A Word Problem a Day*. Palo Alto, Calif.: Creative Publications, 1982.

This is a collection of 180 word problems in a problem-of-the-day format. Weekly challenge problems are also included.

Dolan, Dan, and Williamson, James. *Teaching Problem Solving Strategies*. Menlo Park, Calif.: Addison-Wesley Publishing Co., 1983.

This book is designed for students in grades 6–10. It includes a classroom-tested program designed to teach students the strategies guess and check, make a table, look for a pattern, make a model, eliminate, and simplify the problem.

Goodnow, Judy, and Shirley Hoogeboom. *The Problem Solver* (Books 1–6). Palo Alto, Calif.: Creative Publications, 1987.

A collection of grade-level binders with activities designed to develop a variety of problem-solving skills and strategies. Reproducible activity sheets and teacher notes with detailed solutions are provided.

Greenes, Carole, and others. *Problem-mathics: Mathematics Challenge Problems with Solution Strategies*. Palo Alto, Calif.: Creative Publications, 1977.

Challenging problems for students in grades 7–12, with complete solutions and problem extensions, are included.

———. *Successful Problem Solving Techniques*. Palo Alto, Calif.: Creative Publications, 1977.

Nine basic techniques for solving problems are introduced in this book, written for both students and teachers in grades 6–12.

———. *REACH*. Palo Alto, Calif.: Dale Seymour Publishing Co., 1984.

Separate books for grades 2–8 provide unusual problems for talented mathematics students. Each book is related to a single theme. Problems require finding information in displays; organizing data in lists, tables, and time lines; using deductive logic; guessing and checking; and performing a variety of operations.

———. *Techniques of Problem Solving*. Palo Alto, Calif.: Dale Seymour Publishing Co., 1983.

A box of 200 problem cards is available for each of grades 2–12. Required strategies include guess and check, using resources, choosing the operation, organizing data, and using logic.

Problem-solving workbooks focusing on each strategy are also available. Resource books are available for each of grades K–2. Each book contains thirteen problem-solving activities.

———— . *Techniques of Problem Solving: Beginning Problem Solving (Grades K–2)*. Palo Alto, Calif.: Creative Publishing, 1983.

Each of four resource books contains thirteen problem-solving activities, including a read-aloud, solve-as-you-go story, shorter story-problem activities, and a problem-solving game.

Holden, Linda. *Math on the Wall: Problem Solving Projects for Bulletin Boards*. Palo Alto, Calif.: Creative Publications, 1987.

Each of these three books contains twenty problem-solving activities involving numbers and operations, measurement and geometry, patterns, graphing, and probability. Problem solutions become part of bulletin boards. Teaching instructions include illustrations of bulletin boards.

———— . *Thinker Tasks—Attributes and Logic, Number Patterns, Visual Perception*. Palo Alto, Calif.: Creative Publications, 1986.

This series of three books provides a variety of problem-solving experiences across the grades. Problems are grouped within the three major categories according to type and difficulty.

Krulik, Stephen, and Jesse Rudnick. *Problem Solving: A Handbook for Teachers*. Boston: Allyn & Bacon, 1980.

This book discusses the teaching of problem solving at all levels. Sample problems and activities are included.

———— . *A Sourcebook for Teaching Problem Solving*. Boston: Allyn & Bacon, 1984.

This book, intended for teachers of all grade levels, contains reproducible activities designed to develop problem-solving skills.

Lane County Mathematics Project. *Problem Solving in Mathematics, Grades 4–9* (six books). Palo Alto, Calif.: Dale Seymour Publishing Co., 1983.

Each book provides a variety of lessons dealing with problem-solving skills and strategies. Each student lesson, which can be integrated with the basal text, is supported by a teacher commentary.

Mason, John, and others. *Thinking Mathematically*. Menlo Park, Calif.: Addison-Wesley Publishing Co., 1982.

This book, for the adult learner who wants to improve his or her problem-solving abilities, contains specific ideas for improving performance as well as practice opportunities.

Meyer, Carol, and Tom Sallee. *Make It Simpler*. Menlo Park, Calif.: Addison-Wesley Publishing Co., 1983.

Sixty problems with easy and more difficult variations are provided for students in the middle and junior high school grades. These experiences are designed for small-group work and last approximately fifteen minutes each. Strategies for understanding and solving problems are introduced.

National Council of Teachers of Mathematics. *Arithmetic Teacher*. [Focus issue on teaching problem solving, February 1982.] Reston, Va.: The Council, 1982.

Ideas for teaching problem solving at all levels, K–8.

———— . *Problem Solving in School Mathematics* (1980 Yearbook). Reston, Va.: The Council, 1980.

An analysis of problem-solving issues and some practical suggestions for problem-solving instruction.

O'Daffer, Phares. "Problem Solving: Tips for Teachers." *Arithmetic Teacher*, September 1984 through May 1985 issues.

Special tips for implementing a problem-solving program. Discusses strategies, classroom climate, and ideas to try, and gives additional problems and references on problem solving.

Ohio Department of Education. *Problem Solving—Becoming a Better Problem Solver* and *Problem Solving—a Resource for Problem Solving*. Palo Alto, Calif.: Dale Seymour Publishing Co., n.d.

A discussion of problem solving in the elementary and junior high school. Good pedagogical suggestions and numerous problems analyzed according to strategies used in solutions.

Roper, Ann, and Linda Harvey. *The Pattern Factory: Elementary Problem Solving through Patterning*. Palo Alto, Calif.: Creative Publications, 1980.

This book is designed for grades 2–6. Activities include work with color cubes, pattern blocks, peg boards, and Cuisenaire rods.

Souviney, Randall. *Solving Problems Kids Care About*. Palo Alto, Calif.: Dale Seymour Publishing Co., 1981.

This book contains thirty-four reproducible problem sheets designed for grades 4–8. Suggestions for guiding students through each experience are provided along with background notes and teaching strategies.

Stofac, Valerie, and Anne Wesely. *Logic Problems for Primary People (Grades 1–3)*. Palo Alto, Calif.: Creative Publications, 1987.

Jungle-theme stories contain problems solvable using logic and manipulative aids. Each book includes techniques for organizing and recording data.